COLOUR CENTRES AND IMPERFECTIONS IN INSULATORS AND SEMICONDUCTORS

COLOUR CENTRES AND IMPERFECTIONS IN INSULATORS AND SEMICONDUCTORS

P. D. TOWNSEND
B.Sc., Ph.D.
School of Mathematical and Physical Sciences
University of Sussex

J. C. KELLY
B.Sc., Ph.D.
Associate Professor of Physics,
University of New South Wales

CRANE, RUSSAK & COMPANY, Inc.
NEW YORK

Published in the United States by:
Crane, Russak & Company, Inc.
52 Vanderbilt Avenue
New York, N.Y. 10017

Library of Congress Catalog Card No. 73–76960

© P. D. Townsend and J. C. Kelly 1973

ISBN 0–8448–0209–3

Printed in Great Britain

CONTENTS

PREFACE

With crystals, as with people, it is not their perfections which we find most interesting. It is frequently the nature and number of their defects which determine their usefulness and many of their important properties. Properties as diverse as mechanical strength or colour are controlled by the defects in the solid, and an understanding of the role of the imperfections and the ways in which they can be introduced into the crystal is essential for the preparation of materials. The literature on the subject is enormous. Why therefore have we chosen to add yet another book to it? Because of the growing interest in defects both for their intrinsic physical value and their dominant effect on devices and processes, more undergraduate courses are including material of this nature. Some study of the subject is essential for all graduate students with an interest in the solid state. We have given courses of lectures at both final year undergraduate level and as part of the compulsory graduate courses for higher degree students at both the Universities of Sussex and New South Wales and have not found a book which provides for these needs. At the risk of falling between two stools we have therefore attempted to write a book which starts at a level comprehensible to final year students and yet provides, for more advanced workers, a useful summary of the current state of research and development in both the understanding of the physical processes involved in defect formation and the application of defects to devices. We have in a number of places hazarded an opinion on the way research and applications will develop in the future.

The history of the subject is well documented elsewhere and we have therefore chosen an approach based on the central features that are now well established and the device applications which should be most fruitful in the future, rather than the order in which they were discovered or developed. Thus magnetic properties appear ahead of optical observations because the more recent magnetic measurements have been the most important in unambiguously identifying defects.

We are indebted to many people for helpful and illuminating discussion and to the Imperial Relations Trust, the University of New South Wales, and the Science Research Council for the provision of funds which enabled us to work together at both the Universities of

Sussex and New South Wales. Our thanks are also due to Linda Lammiman for typing numerous drafts of the manuscript and to Douglas Meyer for preparing the figures. We are also indebted to our colleagues and students who criticized the contents of our lecture courses.

1

INTRODUCTION

1.1 Introduction

Writers of undergraduate text books on solid state physics generally choose idealized situations that can be handled completely within the framework of the theoretical models. Whilst this is undoubtedly satisfying for a student, the approach of only considering analytically soluble problems often generates the feeling that applied physics is somehow less challenging because one must frequently resort to an empirical approach if progress is to be made. Even worse is the realization that many industrial processes, such as phosphor preparation or photography, developed initially without any understanding of the basic physical processes involved and even now this understanding is incomplete. The reason for most of these problems is that materials are not perfect but contain a host of defects throughout their structure. Although the number of atomic sites which are affected by the defects may be a small fraction of the total number, the defect behaviour can control the properties of the bulk material.

Possible types of defect, their identification and their properties may receive a brief mention in undergraduate books, but there is clearly a need to expand this material to make the transition from ideal to real physics more acceptable. In so doing one may choose to call the field solid state physics, solid state chemistry or materials science depending on personal background. No matter which choice is made the essential point is that we are considering substances as they occur in reality and not as they occur in idealized models of solids made of repeated identical perfect unit cells. Our aims in this book are to show that useful defect studies are feasible no matter how complex is the system involved and also to show that imagination, intuition and empiricism are an acceptable and necessary combination for the development of new materials and even for the understanding of many old materials.

The importance of crystalline imperfections was appreciated as early as the 1930's when some excellent work was done on colour centres in alkali halides and the mechanism of the photographic process. However, the real impetus in the field did not occur until after 1945 with the subsequent development of reliable commercial optical and electron spin resonance (ESR) spectrometers, electron microscopes and similar instruments. Improvements in crystal preparation have produced specimens with reproducible impurities and defects, and this reduction in the influence of random effects has caused a tremendous proliferation of defect studies. In part this rapidly changing scene has been possible because of the continuous interplay between empiricism and basic understanding. Certainly the semiconductor industry has many examples which demonstrate that detailed defect studies followed, rather than preceded, material development, whilst in other fields the development of devices had to await the solution of basic problems.

A further result of introducing only simple systems into undergraduate teaching is that one becomes conditioned to think in terms of simple systems. For example chemical composition is always mentioned in terms of Dalton's law and deviations from this law are treated as special cases of nonstoichiometry. It is, however, now clear that systems such as Fe_xO_y, Ti_xO_y, V_xO_y exist over a wide range of values of x and y and form regular lattices in which defect sites are an integral part of the array. Defect studies in such materials are correspondingly more complex than studies in alkali halides or silicon, but the problems are still tractable.

1.2 Simple defects

Since we view this book as a logical extension of normal undergraduate solid state physics we shall assume that the reader is familiar with some crystallography and has been introduced to quantum mechanics and of course developed this to the basic concepts of band theory.

The smallest defects in a crystal structure that can be used as building blocks for complex defects are the point or atomic defects of vacancy, interstitial or impurity atom. The basic extended defects are dislocations and surfaces, both internally at cracks and voids as well as at the outer surface. Vacant lattice sites can occur in all materials, but in insulators or semiconductors the empty site can exist in many charge states. Similarly interstitial atoms occur in a variety of charge states, and the stable configuration for an interstitial may depend on this charge, as for example the crowdion halogen interstitial in the alkali halides (see § 7.3). Interstitial atoms can be accommodated in several ways into the lattice and three possible models are shown in

Fig. 1.1. These are a body-centred site, a pair af atoms sharing one lattice site in a dumb-bell fashion and a crowdion interstitial where four atoms share three linear lattice sites. Impurity atoms can enter the lattice either substitutionally, interstitially or in any complex with other defects. From the simple band picture of an insulator we expect that the localized disturbance of the lattice will produce additional electron energy levels which may fall in the normally forbidden energy region between the valence and conduction band. It is the occupation of, and transitions between, these levels that produces all the interesting defect-controlled electrical and optical properties of the materials we wish to study. When many levels exist between the

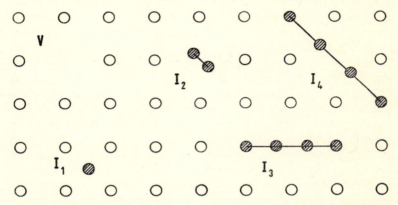

Fig. 1.1 Simple point defects in a lattice. V is a vacant lattice site. Interstitial sites shown are I_1, a body-centred site; I_2, a dumb-bell pair; I_3 and I_4, crowdion sites.

valence and conduction band the electron population of these localized levels is determined by the position of the Fermi level. Those below the Fermi level will in general be occupied, those more than a few kT above will be empty, under equilibrium conditions.

For much of this book we will be concerned primarily with point defects, but materials also have line defects. Crystals typically contain some 10^6 dislocations per square centimetre, and their presence influences both mechanical properties and the rate at which point defects are generated under energetic irradiation. For example dislocations reduce the mechanical strength of materials, compared with that of a perfect lattice, by a factor of 10^4. Whiskers, which are dislocation free, come nearest to the ultimate tensile strength of a material. As we will discuss in Chapter 6 dislocations in non-metals may be charged with respect to the perfect lattice, and this together with the strain field around the dislocation produces concentration gradients of defects throughout the material. A similar effect is produced by surfaces, which in real problems cannot be removed by the application of cyclic boundary conditions. In these cases the electric fields separate free electrons and holes to different parts of the crystal. In small crystallites, such as the silver halide grains used in photographic emulsions, this charge separation is a key step in making the photographic process operative.

1.3 Theoretical treatments

Considerable efforts have been made to use colour centres as model systems for quantum mechanical calculations. For the simpler centres the calculations of transition energies produce results within 10 or 20% of the observed energies, but for better agreement between theory and experiment one often resorts to fitting of scaling parameters. This has some value in deciding the properties of identified defects, but is not helpful for unidentified centres. For example the first transition energy for the vacancy in diamond is variously estimated between 1 and 8 eV. This is a particularly bad case, but it does indicate that one is unlikely to distinguish between defects with similar energy level spacings. For this reason we shall not discuss the theoretical calculations of the energy levels of defects. Some specific references will be given for particular centres, and the general treatments are also mentioned in the following references to background literature.

General References

Brown, F. C., *The Physics of Solids* (Benjamin) 1967.

Compton, J. H. and Schulman, W. D., *Color Centers in Solids* (Pergamon) 1963.

Fowler, W. B., editor *Physics of Color Centers* (Academic Press) 1968.

Greenwood, N. N., *Ionic Crystals, Lattice Defects and Nonstoichiometry* (Butterworth) 1968.

Hannay, N. B., *Solid State Chemistry* (Prentice-Hall) 1967.

Kittel, C., *Introduction to Solid State Physics* (Wiley) 1971.

Mott, N. F. and Gurney, R. W., *Electronic Processes in Ionic Crystals* (Dover) 1964.

Wert, C. A. and Thomson, R. M., *Physics of Solids* (McGraw-Hill) 1964.

2

MAGNETIC PROPERTIES

2.1 Electron spin resonance – basic theory

Of all the experimental techniques applied to studies of defects in insulators electron spin resonance (ESR) is the most powerful. Successful interpretation of the spectra not only provides the basic structure of the defect in the ground state, but also indicates the distribution of the charge over the surrounding nuclei. If one can also make electron nuclear double resonance (ENDOR) measurements one can estimate the interactions over several nearest neighbour shells of atoms, in favourable cases this has been possible even as far as the thirteenth shell. When one realizes that this means that a simple point defect is interacting with some 2000 surrounding atoms, it is evident that extremely sensitive techniques are required for studying defects. An unfortunate consequence is that theoretical calculations of the energy levels around defects are too complex to yield the accuracy obtainable by experiment.

The detailed theory of electron spin resonance has been given in many places [1, 2, 3] and the applications of ESR and ENDOR to F centres in alkali halides has been excellently reviewed by Seidel and Wolf[4].

The following introduction to this field will demonstrate the power of this tool and the type of information that it can yield. Reference will also be made to the experimental procedures, since the measurements themselves may alter the observed spectra. Even this may be employed to advantage if one has samples with several defects with overlapping spectra, since one may be able to saturate preferentially some resonances.

In ESR one is studying a defect by observing the splittings between electron energy levels which are produced by interaction of the defect with crystalline fields, nuclear spins and the external magnetic field. The simplest case one can consider is a free ion with one electron in an energy level with a spin state $S = \frac{1}{2}$. An applied magnetic field H will remove the degeneracy and split the energy level into two levels given by the spin magnetic quantum numbers $m_s = \pm\frac{1}{2}$. The magnitude of the splitting is $\Delta E = g\mu_B H$ where g is the spectroscopic splitting factor and μ_B is the Bohr magneton $(\mu_B = \dfrac{e\hbar}{2mc})$. Transitions will occur in the presence of an electromagnetic field at the resonant energy $\Delta E = h\nu$. It is thus possible to scan through the spectrum of energy levels by changing either the magnetic field H or the frequency ν. The allowed transitions obey the selection rule $\Delta m_s = \pm 1$.

The electromagnetic waves are polarized and have a magnetic field strength H_1 which precesses about the stationary magnetic field H_0 as shown in Fig. 2.1. (In classical magnetic terms we would

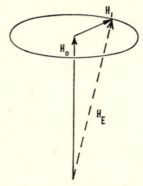

Fig. 2.1 The relation between the static magnetic field H_0 and the precessing microwave field H_1.

have said the magnetic field of the radiation provided the torque to change the state of the magnetic ion or defect.) The effective field at the defect at any instant is the vector sum of the two applied fields.

In Fig. 2.2(a) and 2.2(b) we consider the number of resonance lines that we expect to find (a) for a degenerate set of levels split by an increasing magnetic field with a fixed frequency electromagnetic field if $s = \frac{1}{2}$ and (b) for the same pair of levels which are further split by a nuclear interaction of $I = \dfrac{3}{2}$. In the second case we must also observe the selection rule $\Delta I = 0$.

It is clear that we are more likely to resolve the detailed structure at higher magnetic fields. Under laboratory conditions uniform fields of 10^3 to 10^4 gauss can be conveniently produced. The frequency of the electromagnetic wave is related to the field H_0 at resonance by

$$gH_0 = 7{\cdot}14 \times 10^{-7}\nu.$$

For a free electron the spectroscopic splitting factor takes the value of $g = 2$ so that at a frequency of 10^{10} Hz we require a field of 3570 gauss. Since this frequency corresponds to the microwave region of the spectrum with a 3 cm wavelength it is most convenient to use a fixed frequency and vary the magnetic field.

If we consider an electron bound to some site within a lattice it will

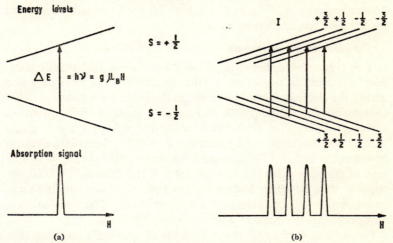

Fig. 2.2 The microwave absorption pattern expected from (a) a spin $S = \frac{1}{2}$ pair of energy levels and (b) the hyperfine line pattern produced by interaction with a nucleus of $I = 3/2$.

have localized energy levels which are similiar to those of a free electron with $g = 2$. But more subtle changes are produced by the environment, and g will become a tensor as the crystal field shapes the electronic orbit. In order to detect this we apply an external magnetic field and further distort the orbit because of the extra interaction. It is thus essential to consider the spatial relationship of the defect with both the crystalline fields and the external magnetic field.

The spin Hamiltonian

In order to describe the energy levels of a particular defect in all their detail it is customary to write a Hamiltonian $\mathcal{H}$ in which we separate the energy contributions of the various types of interaction. However, it is simpler to understand the origin of the various terms than actually perform the calculation in a real case, also individual problems may justify a Hamiltonian cast in quite a different form. Since we are dealing with electrons in a quasi-free environment the electron Zeeman term ($g\mu_B H.S$) is still dominant and the other terms can be treated by perturbation theory. Thus we write

$$\mathcal{H} = \mathcal{H}_0 + \mathcal{H}_1 + \mathcal{H}_2 \cdots$$

where the terms represented by the interactions are $\mathcal{H}$ = electron Zeeman ($\mathcal{H}_0$) + electron Stark ($\mathcal{H}_1$) + electron nuclear ($\mathcal{H}_2$) + nuclear Zeeman ($\mathcal{H}_3$) + nuclear quadrupole ($\mathcal{H}_4$) . . . . We shall now examine each of these in turn, both for the form of the term and for the information that it can yield.

The electron Zeeman term

This is the zero order term of the Hamiltonian and therefore defines the central transition of the resonances. As we mentioned earlier the electron is not free and is in the field of a number of other charged atoms. Consequently the scalar g value is replaced with a tensor $\tilde{g}$ which differs from the free electron value by $\Delta g = g_{bound} - g_{free}$. The magnitude of Δg describes how tightly bound the electron is within the defect. Clearly a spacious defect will have $\Delta g \approx 0$. The sign of Δg also gives the charge state of the defect. For ESR we required that the centre had an unpaired spin, and this condition can be satisfied by either an electron or a positive hole. The electron gives a negative Δg, the positive hole a positive Δg.

Finally one determines the magnitude of spin orbit coupling that the electron (or hole) has with the lattice by the tensor components of $\tilde{g}$. Thus we expect the first term to be $\mathcal{H}_0 = \mathcal{H}.\tilde{g}\mu_B S$.

The electron Stark term

From classical atomic physics one might expect the largest perturbation of the energy levels to come from the electric field interactions (i.e. fine structure). This is not always the case for electrons bound to defects, and it may not be much larger than the electron-nuclear term (i.e. hyperfine structure) in some materials. The Stark splitting of the levels comes from a spin-spin coupling via the orbital terms. Consequently the term involves a tensor coupling $\tilde{D}$ between two spins so a general form of it may be $S.\tilde{D}S$. The magnitude of the splitting of the energy levels depends on the applied magnetic field. In the treatment by Bleaney[5] the major term is of the form DS_z^2 $(3\cos^2\theta - 1)$. Rotation of the magnetic field around the crystal thus alters the spacing of the lines. The apparent number of lines also alters in certain crystalline directions where the energy levels are degenerate.

For example an ESR study of defects in an electron-irradiated diamond by Faulkner and Lomer[6] revealed three separable defects with different crystalline axes. The line pattern could be separated into pairs of lines centred on the free electron resonance for $g = 2$. In Fig. 2.3(a) one has simplified the angular variation pattern to a six-

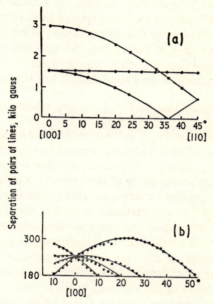

Fig. 2.3(a) Angular variation of pairs of absorption lines found in diamond as a function of the angle between H_0 and the [100] direction in the (001) plane. The curves were calculated for $S = 1$, $D = 0.138$ cm^{-1}, and a z-axis along $\langle 100 \rangle$.

Fig. 2.3(b) A second set of lines viewed with H_0 varying from [100] in the (011) plane. Theoretical curves are for $S = 1$, $D = 0.0141$ cm^{-1} and a z-axis of $\langle 932 \rangle$.

line spectrum by aligning the magnetic field with the $\langle 100 \rangle$ crystalline direction. One should note the very large separation of these lines. For comparison we see in Fig. 2.3(b) the strongest lines of a more complex and compact line pattern which was also found in the diamond. Theoretical line variations are shown for the values $S = 1$, $D = 0.138$ cm^{-1} and a z-axis in the $\langle 100 \rangle$ direction for Fig. 2.3(a). For the calculated curves of Fig. 2.3(b) a high order axis of $\langle 932 \rangle$ is used with an $S = 1$ and $D = 0.0141$ cm^{-1}.

The value of D can be readily derived from the angular pattern as the $(3 \cos^2 \theta - 1)$ term oscillates between $+2$ and -1 when the field is parallel to the defect axis, and the total line shift thus gives DS_z^2.

The electron nuclear interaction

Electron energy levels may be split by nuclear interactions into $(2I + 1)$ levels if the nucleus has a spin quantum number I, but such a splitting is only possible for electron levels which have a finite wave function at the region of the nucleus. An alternative name for this is the contact interaction. As before we may describe the strength of the coupling by a tensor $\tilde{A}$, in a term $I.\tilde{A}.S$, but here we should realize

that A is proportional to $\dfrac{\mu_N \mu_B}{I} |\psi_{(0)}|^2$, where μ_N is the nuclear

magnetic moment and $|\psi_{(0)}|^2$ is the probability of finding the electron at the nucleus. Therefore if we replace one nucleus by a different one both I and A change, and we alter both the number of hyperfine structure lines $(2I + 1)$ and the spacing of the lines which is proportional to A. The spacing of hyperfine structure absorption lines is constant between all the lines at any particular setting of the magnetic field with respect to the defect axis. We can thus separate hyperfine structure lines and crystalline field lines by rotating the sample.

The appearance of electron nuclear interactions is particularly powerful in the identification of a defect because we can change the isotopic abundance of the elements in the crystal.

For example, consider a simple defect which involves an electron and a nearby nucleus of potassium. Potassium has three separable isotopes K^{39}, K^{40}, K^{41} with spins $I = 3/2$, 0 and 3/2. The coupling constant A for K^{39} is 0·007 cm^{-1} and for K^{41} is 0·0042 cm^{-1}, derived from the experimentally observed values $\mu_N = 0·261$ for K^{39} and $\mu_N = 0·143$ for K^{41}. If the electron is only bound to one potassium ion we expect a $(2I + 1)$ hyperfine line spectrum. For these three isotopes of K^{39}, K^{40}, K^{41} this means four-line, one-line or four-line spectra. In addition the coupling constants for the K^{39} and K^{41} spectra are in the ratio 700 : 424 so this will also be the ratio of the total widths of the two sets of hyperfine lines.

It was the use of this approach by Hutchinson and Noble[7] that finally confirmed the de Boer model of the F centre in alkali halides (a detailed list of the established models for defects in alkali halides is given in Chapter 7). This very famous centre is a halogen vacancy which has been stabilized by the addition of an electron. The centre thus retains the original symmetry of the halide ion and interacts equally with six nearest neighbour metal ions. In KCl each K^{39} has a nuclear spin of 3/2 which raises the total possible spin to $I = 9$. The number of hyperfine lines in the spectrum is raised to 19. Since there are six neighbours in a particular spin state the probability of having all the spins aligned is a minimum and can only be achieved one way for $I_{total} = 9$. However, the $I_{total} = 0$ case is possible for 580 arrangements of the nuclear spins. Clearly the components of the 19 line spectra are not all equally intense and a summation of the possibilities gives weighting factors shown by Fig. 2.4.

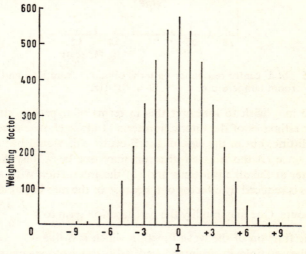

Fig. 2.4   The statistical weights of the hyperfine resonance lines produced by interaction with six equivalent nuclei of $I = 3/2$.

If the electron nuclear interactions is strong this eases indentification of the defect since, in this case, all 19 lines would appear in the spectrum, and this severely limits the number of possible models for the centre. In practice such a large number of hyperfine lines may overlap, and the weighting factors produce a single broad resonance with approximately a Gaussian envelope. Examples of the F centre resonance are shown for KCl and NaF in Fig. 2.5. One displays the spectrum as the first derivative of the absorption curve since this emphasizes the number of components. It will be seen that with the viewing conditions used for KCl the hyperfine structure is not apparent. The line is extremely broad for a single absorption line, and under some conditions hyperfine structure may be detected.

In resonance conditions where we observe a very broad resonance

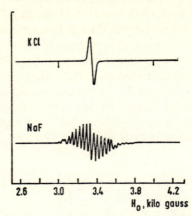

Fig. 2.5 The F centre resonance patterns obtained from KCl and NaF at room temperature. $\nu = 9.38 \times 10^9$ Hz.

line we may wish to interpret this in terms of hyperfine structure. Earlier estimates of the interaction term $\tilde{A}$ alone should predict the line splitting, but in the case of the F centre this would be inappropriate since (i) the trapped electron may not be wholly S-like in character at the alkali nucleus and (ii) the interaction with a single nucleus is reduced by a factor of 6 because of the number of identical neighbours. One may modify the perturbation term to $Y \dfrac{I.\tilde{A}S}{6}$ where Y is the fraction of the wave function which is S-like.

Further complexity is introduced if several isotopes exist in the sample, since nuclear neighbours will vary through all possible combinations of nuclear spin and $\tilde{A}$ according to the isotopic abundance.

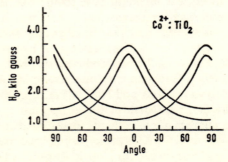

Fig. 2.6 Variation of some hyperfine structure lines of Co^{2+} in TiO_2 at 4·2K as the sample is rotated in the ab plane of the crystal.

At first sight the presence of hyperfine structure produces a complex spectrum, but once we have a model for a defect it enables us to make a positive identification.

As a simple example we can consider the hyperfine structure lines which result from the addition of Cobalt^{2+} into a single crystal of

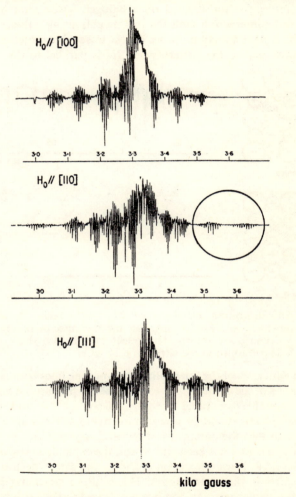

Fig. 2.7 The dependence on crystalline direction of the hyperfine structure of the H centre in KCl. Measurements made at 20 K and $9 \cdot 277 \times 10^9$ Hz by Kanzig and Woodruff. The circled region is shown in detail in Fig. 2.8.

rutile[8]. In Fig. 2.6 we see that there are two pairs of lines with a separation which is only slightly modified by rotation of the magnetic field. The equal spacing of the hyperfine lines is superbly demonstrated in the paper by Kanzig and Woodruff[9] on the electronic structure of the H centre in alkali halides. This centre is basically a halogen molecular ion located on a halogen lattice site; however, there is some interaction with the two neighbouring halogens along a <110> crystal direction. So we expect to see fine structure with a minimum spread to the pattern when H_0 is parallel to the <110>

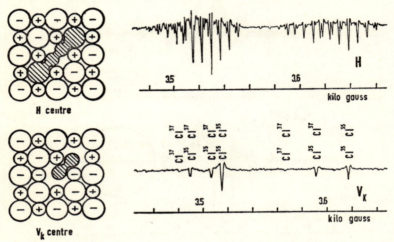

Fig. 2.8 An illustration of the resolution obtainable in the ESR measurement. For comparison the line spectra of the H and V_K centres are shown. The defect models also show how the hole is trapped in the lattice in each case.

direction with each of the lines split into hyperfine lines because of the nuclear spins of the chlorine. Three views of the pattern are presented in Fig. 2.7 with H_0 parallel to the <100>, <110> and <111> crystal axes. Unfortunately one cannot avoid the strong F centre resonance so that the pattern as obtained is asymmetric.

The side structure has been circled and the expanded view of this is seen in Fig. 2.8. For comparison is shown the same spectral region for V_K centres in the alkali halides. This is also a molecular halogen ion with <110> symmetry, but it is not bound to the extra halogens that occur in the H centre. We should remember that chlorine has two isotopes Cl^{35} and Cl^{37} in abundances of 3 : 1. For a molecular ion combinations of $Cl^{35}-Cl^{35}$, $Cl^{35}-Cl^{37}$ and $Cl^{37}-Cl^{37}$ can occur. With the addition of two more halogens in the H centre the hyperfine structure is multiplied by these additional nuclear spin interactions.

In an example such as this where one can proceed from the simple V_K centre to the similar H centre it is obvious that there is justification in identifying the origin of each resonance line.

In summary we see that a defect which has an ESR showing hyperfine structure can be tested against quite detailed models of the centre. Although the picture is complex even more nuclear interactions can be resolved by ENDOR.

Small perturbations

For simple ESR the remaining perturbations are small and may be neglected, but in the double resonance experiment (ENDOR) they are important, as they may produce relatively strong resonances despite the fact that they are 'forbidden' transitions. The nuclear Zeeman term is $H\tilde{g}_N\mu_N.I$ and has the selection rules $\Delta m_s = \pm 1$, and $\Delta m_I = \pm 1$. Another term involving nuclear quadrupole interactions $I.\tilde{Q}.I$ may be confused with the nuclear Zeeman term.

In the above we have assumed that the interaction was curtailed after the nearest neighbour. This is not true and each of these energy levels is further broadened by hyperfine coupling to more distant nuclei. We have also assumed that all the interactions take place about the same symmetry axis. We are justified in doing this since a variety of axes does not change the type of interactions. However, it does add considerably to the mathematics.

Summary of the basic ESR information

The ESR spectrum gives information on the charge state of a defect, the type of binding, the amount of interaction with nearby nuclei and the symmetry of the environment. This is sufficient information to propose a fairly detailed model of the centre. Up to this point we have only predicted the number of energy levels available for the electron without any regard to the difficulty or conditions required for viewing such transitions. The resonance condition requires that power be absorbed from the microwave beam at a constant rate. Thus we must allow the excited state to lose energy either by stimulated downward transitions or by some spin-spin coupling to the lattice or other defects. If not, we merely produce population inversion, and the resonance signal is reduced in intensity when all the electrons are raised to the higher state. This means that low power levels are desirable. However, for a good signal to noise condition in the electronics we want high power levels. We will therefore consider some of the experimental arrangements and their relationship to the measured signal.

Experimental procedure

ESR spectrometers may be quite sophisticated electronic systems and we shall not describe them in detail. A full treatment of the subject is given by Ingram[10]. For our purposes the problem is to measure power absorption from the microwave beam. It is essential to realize that the spectrometer is not a passive element, and the detection system modifies both the line shape and the line intensity. Simple transmission experiments are poor since we are looking for small changes in the presence of a large signal. Classical microwave improvements on this are (i) to use a resonant microwave cavity to increase the microwave field H_1; (ii) to lock the klystron to the resonant frequency of the cavity, even if this changes as one passes through the sample resonance and alters the Q of the cavity; (iii) to modulate the signal by a small A.C. magnetic field imposed on the slowly changing H_0, and then to use a tuned detector circuit together with phase sensitive detection. If these are still inadequate, one can improve the signal to noise ratio by repeatedly running the same spectrum and summing the spectra on a multichannel analyser. The noise being random it should average towards zero whereas the signal components will add. In practice the improvement in the signal to noise increases as the square root of the number of repetitions. Finally it is customary to display this signal as the first derivative of the absorption curve observed as one sweeps through the magnetic field. The advantage of this is that one emphasizes the small nuances of the absorption spectrum which hide a number of overlapping peaks.

Problems arise because the electronic requirements for sensitive detection may not be compatible with the sample requirements and for most defects one requires a spin concentration of 10^{15} spins cm^{-3}. This concentration may seem high because a good spectrometer should have a minimum detection sensitivity of 10^{11} spins. However, the small sample size (say a 3 mm cube to avoid overloading the resonance conditions of the cavity) and problems of line breadth and stray impurity lines increase the minimum working concentration of defects to 10^{15} spins cm^{-3}.

A more complex problem involves the way in which the energy is dissipated from the defect after it has been stimulated to the higher spin state. Relaxation may occur by direct interaction with the lattice. This is called spin-lattice relaxation and is characterized by a relaxation time T_1. An alternative mechanism is a spin-spin interaction with other defects. This has a relaxation time T_2. Since T_1 requires thermalization with the lattice, it is much longer than T_2 and is also temperature dependent. For example in KCl the spin-lattice relaxation time T_1 varies from 10^{-4} s at 300 K to 100 s at 2K whereas the spin-spin relaxation T_2 remains close to 10^{-6} s.

The differences between the two types of relaxation also influence

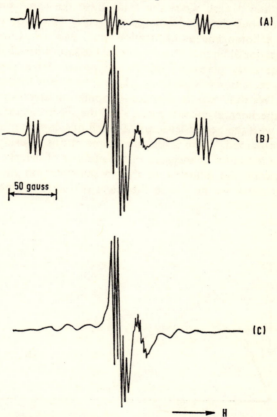

Fig. 2.9 The effect of microwave power on the ESR signal measured in a sample of KN_3 which contains gamma ray induced defects. T = 77 K with H_0 along [001], $\nu = 9.2 \times 10^9$ Hz. The power levels are (A) 0.002 mW; (B) 0.2 mW; (C) 20 mW.

(i) the line shape; (ii) the power saturation curve and (iii) the response of the overall shapes to intensity increases at a frequency within the absorption line.

Since the spin lattice relaxation is a random process with the lattice spins having different precessional frequencies it is called an inhomogeneous process. The result is an absorption band which is roughly Gaussian in shape. Secondly, the lattice provides a large reservoir of spins which exceeds the number of defect spins, so we may increase the microwave power and thus increase the absorption signal up to an equilibrium value. In contrast the relaxation via spin-spin coupling is fast, since the spins have the same precessional frequencies, but there is no large reservoir. This type of process produces a homogeneous absorption band which is Lorentzian in shape. A major difference also appears in the saturation characteristic of the line with increasing microwave power. High power levels produce a reduction in signal since there is no energy loss mechanism available. We thus rapidly produce a population inversion. This has serious experimental consequences, since the spectrometer sensitivity increases with the power level and a small defect concentration can easily be obscured. There is also no simple means of using the ESR signal as a quantitative measure of the number of defects.

To illustrate the influence of the spectrometer on the observed resonance signal we may note the effect of increasing power on a

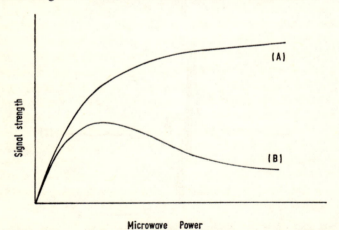

Fig. 2.10 The behaviour of the ESR absorption signal as a function of microwave power for (A) an inhomogeneously broadened line and (B) a homogeneously broadened line.

signal in potassium azide[11]. In this example, shown in Fig. 2.9, the spectrum was taken with a crystal at 77 K with the field H_0 along the <001> direction. In this example the power levels used were 0·002 mW, 0·2 mW and 20 mW and, as is evident from the figure, these power levels produce such large changes that we would not suspect curves A and C were taken from the same sample.

In some instances changes in the spectra induced by changes in the power level can be of value since we can resolve different overlapping spectra, but it should also act as a warning that a simple measurement does not yield all the information.

The theory of the curve shapes was first considered by Bloch[12]. The many later developments, and limitations, of the theory are discussed in all standard texts of ESR. The absorption signal is the imaginary part of the measured susceptibility χ'' and the two types of saturation curve are shown in Fig. 2.10. The real part of the susceptibility χ' is the dispersion signal. (We can choose the type of signal by the relative phases of the exciting and detected signal.) Since the dispersion signal does not saturate, this is frequently studied for inhomogeneously broadened lines and in electron nuclear double resonance.

2.2 ENDOR Electron nuclear double resonance

When considering the interaction of a trapped electron with its neighbouring atoms we determined the number of ESR lines by considering only nearest neighbour interactions. In the case of an F centre in KCl we correctly predicted a 19-line spectrum. However, we also realize that the electron density is finite beyond the first neighbours so we should look for the hyperfine interaction terms from more distant nuclei. Our apparent success in KCl for the F centre has shown that the electron is mostly bound within the six nearest neighbours, but the line width of each of the 19 lines is broadened by the longer range interactions. Analysis of the overlapping components of each of the ESR lines will give even greater detail about the weaker interactions (i.e. coupling strengths and symmetry). This is feasible because the selection rule for nuclear transition of a single nucleus is $\Delta m_I = \pm 1$, but all cross interactions between nuclei have $\Delta m_{I_1 I_2} = 0$.

In the simple case shown in Fig. 2.11 both the energy levels are split by the nuclear-nuclear interactions. However, the two levels are split by different amounts. To resolve this structure we first saturate the major transition say from level A_1 to B_2 with microwave power.

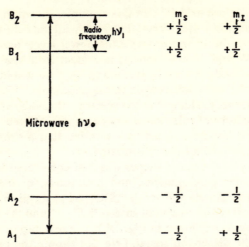

Fig. 2.11 The energy level scheme for an ENDOR experiment.

This produces population inversion in level B_2 compared with level B_1. Whilst in this condition we observe only a weak microwave absorption. However, if we now stimulate transitions between B_2 and B_1 by a radio frequency signal the microwave absorption will re-appear at resonance. We then determine the angular dependence of the spectrum on the axis of the external magnetic field in the usual way.

As an example we shall consider the case of an F centre in CaF_2 which is described by Hayes and Stott[13]. The F centre is a fluorine ion vacancy in which there is a trapped electron. The hyperfine splittings occur from interactions with fluorine nuclei ($I_F = \frac{1}{2}$, $I_{Ca} = 0$). The fluorite lattice is sketched in Fig. 2.12, and from this model we see that the six nearest neighbours to an F type centre all

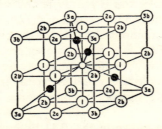

Fig. 2.12 A sketch of the atomic arrangement of an F centre in a fluorite lattice. Calcium atoms are shown as solid circles. Nearest neighbour fluorines are labelled 1, second shell fluorines 2a, etc., third shell atoms 3a, etc.

lie in <100> directions. Consequently we obtain a seven-line ESR spectrum with line intensities in the ratios 1 : 6 : 15 : 20 : 15 : 6 : 1. Even for these ions we may resolve greater structure in the energy levels depending on the relative orientations of the electron spin and the applied field. More obviously the second shell of fluorine ions, labelled 2a, b, c, are directed at <110> to the F centre so that an angular plot of the levels as the field H is rotated in the (01$\bar{1}$) plane gives ENDOR resonances which distinguish between the 2a, 2b and 2c type atoms and show the <110> symmetry. This result is shown in Fig. 2.13. Similarly the third shell of fluorine atoms has <111>

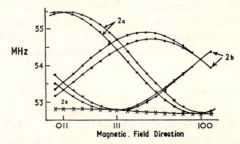

Fig. 2.13 Angular dependence of second shell ENDOR lines in CaF_2. Note the lines from type 2a, 2b and 2c atoms are well resolved.

symmetry. The results shown in Fig. 2.13 clearly separate these interactions.

The sensitivity of the ENDOR measurements is such that it is possible to detect interactions with at least seventh shell neighbours[4] where the electron density of the F centre electron is decreased by 10^6 from the central density. Such quantitative and detailed information provides an excellent testing ground for theoretical calculations, and one is frequently in the position of being limited by the theoretical complexities.

2.3 Other magnetic techniques

Two other magnetic techniques have been used in defect studies, but both have rather low sensitivity and do not provide the detailed information of ESR and ENDOR. The first of these is a study of nuclear magnetic resonance lines in imperfect crystals where the coupling to defects alters the nuclear spin-lattice relaxation time I_1 and the width of the line. The second method is to measure the static paramagnetic susceptibility of a crystal containing defects. Again the sensitivity of the measurements limits the technique to defect concentrations of some 10^{18} defects cm^{-3}. This may be valuable if one merely attempts to decide between two competitive models for a defect which have vastly different magnetic properties. However, no detailed discussion of the methods seems necessary at present.

2.4 Summary

The preceding introduction to ESR and ENDOR has attempted to show that in insulating materials a defect which contains an unpaired electron spin can generally be identified. In particular the interactions with neighbouring atoms, the crystalline and external magnetic fields are frequently sufficient to provide a detailed and unique model for the defect. The discussion has also included some comments on experimental procedures, because this is not a passive measurement and the detection methods influence the spin states and their population.

References Chapter 2

[1] Abragam, A., *The Principles of Nuclear Magnetism* (Oxford), 1961.

[2] Pake, G. E., *Paramagnetic Resonance* (Benjamin), 1962.

[3] Slichter, C. P., *Principles of Magnetic Resonance* (Harper and Row), 1963.

[4] Fowler, W. B., Editor, *Physics of Color Centers* (Academic Press), 1968. Seidel, H, and Wolf, H. C., Chapter 8.

[5] Bleaney, B., *Phil. Mag.* **42**, 441, 1951.

[6] Faulkner, E. A. and Lomer, J. N., *Phil. Mag.* **84**, 1995, 1962.

[7] Hutchinson, C. A. and Noble, G. A., *Phys. Rev.* **87**, 1125, 1952.

[8] Yamaka, E. and Barnes, R. G., *Phys. Rev.* **125**, 1568, 1962.

[9] Kanzig, W. and Woodruff, T. O., *J. Phys. Chem. Solids*, **9**, 70, 1958.

[10] Ingram, D. J. E., *Spectroscopy at Radio and Microwave Frequencies* (Butterworth), 1967.

[11] Shuskus, A. J., Young, C. G., Gilliam, O. R. and Levy, P. W., *J. Chem. Phys.* **33**, 622, 1950.

[12] Bloch, F., *Phys. Rev.* **70**, 496, 1946.

[13] Hayes, W. and Stott, J. P., *Proc. Roy. Soc.* **A301**, 313, 1967.

3

OPTICAL ABSORPTION AND LUMINESCENCE

3.1 Optical absorption and luminescence of localized energy levels

Observations of defects in insulators by the colour changes produced in the crystal is the simplest and most appealing of all the techniques used. The analysis of the data does not have the finality of ENDOR measurements, but the beauty of stained glass, ruby or sapphire has directed many people into this field of study. Since the visible spectrum extends from 1·7 to 3·1 eV, visual colour centres can only occur in wide band gap material. However, the methods of analysis are also relevant for infra-red absorption bands in the narrower band gap materials such as semiconductors.

Photon absorption occurs in a material when a suitably spaced pair of energy levels exist which have an electron or hole in the lower level. For the transition to be allowed one also requires that the upper state is of the reverse parity. Momentum conservation is possible, even if the levels are not vertically displaced on an E, k (energy, momentum) diagram, by addition or subtraction of lattice phonons.

It is the interaction of the phonons with optical transitions which determines the shape of the fundamental absorption edge and the width of absorption and emission bands. For the case of transitions between the valence and conduction bands the two simplest examples are shown in Fig. 3.1. The energy gap shown in Fig. 3.1(a) is in between the maximum of the valence band and the minimum of the conduction band, and since both curves are centred on the zero k line the minimum energy required to promote an electron to the higher band can be provided by a photon without a phonon interaction. This is called a direct transition.

Many systems (e.g. Ge, Si, AgBr) do not have the band minima at $k = 0$. Thus the energy gap may be smallest for a sideways transition involving phonons. These are called indirect transitions and include the cases of both phonon addition or emission to the photon energy. Examples of both cases are shown in Fig. 3.1(b). For

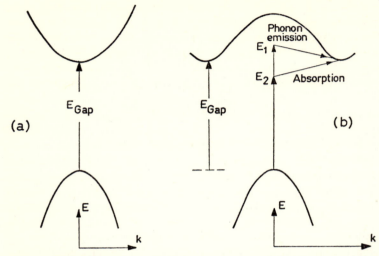

Fig. 3.1 Examples of E, k diagrams which result in optical absorptions by (a) direct transitions and (b) indirect transitions.

momentum conservation the phonon momentum must equal the difference between the momenta of the initial and final electron states, so although a photon of energy E_1 is above the band gap energy a transition is only possible if phonons are emitted, similarly an energy of E_2 can also cause the transition if in addition phonons are absorbed during the transition.

Since the density of filled states at the top of the valence band is high the transition probability is also high for energies around the band gap energy and all materials become opaque at these energies. At very low temperature the absorption edge is almost a step function, but thermal fluctuations in the density of filled levels produce a tail to this edge. Analysis of the tail shape reveals whether the process is a direct transition or involves phonons and whether the two states differ in parity or are normally forbidden[1,2]

On measuring the absorption coefficient μ (discussed in § 3) as a function of the distance from the fundamental edge we find:

$\mu \propto |E_{\text{photon}} - E_{\text{gap}}|^{1/2}$ for direct allowed transitions
$\mu \propto |E_{\text{photon}} - E_{\text{gap}}|^{3/2}$ for direct forbidden transitions
$\mu \propto |E_{\text{photon}} - E_{\text{gap}}|^{2}$ for indirect allowed transitions
$\mu \propto |E_{\text{photon}} - E_{\text{gap}}|^{3}$ for indirect forbidden transitions.

In the case of indirect transitions a plot of $\mu^{1/2}$ versus the photon

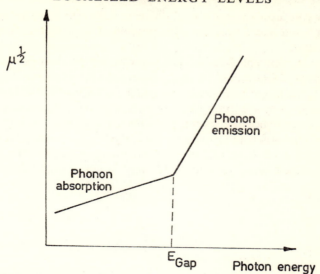

Fig. 3.2(a) An example of the changes in absorption coefficient with photon energy for indirect transitions.

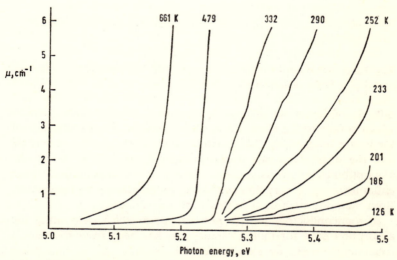

Fig. 3.2(b) Examples of the absorption edge spectrum of type IIb diamond at various temperatures.

energy can show two regions of differing slope which correspond to the emission and absorption of a phonon. This is demonstrated in Fig. 3.2(a).

A large apparent shift in band gap with temperature is shown in the results of Clark *et al.*[3], Fig. 3.2, for type IIb diamond. The type IIb diamonds are sufficiently pure that the absorption spectra are not confused by absorption from extrinsic defects. The sharp changes in slope occur because several phonons are involved. On replotting data of this form with an ordinate scale of $\mu^{1/2}$ or $\mu^{3/2}$ one can readily detect the type of transition involved for each part of the curve. Idealized and actual data for CdTe are shown in Fig. 3.2(c). Clark

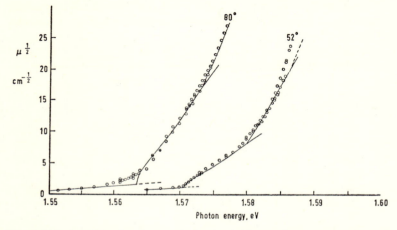

Fig. 3.2(c) An example of the $\mu^{1/2}$ plot for CdTe. (After D. T. F. Marple, *Phys. Rev.* **150**, 728, 1966.)

interpreted his set of data as combinations of six lattice phonons which couple to the photon absorption process. The band gap is of the form shown in Fig. 3.1(b) with an indirect energy gap at 5·47 eV and the direct (vertical) transition energy of 7·02 eV. When the absorption coefficient is so large (100 cm^{-1}) the measurements must either be made on very thin samples or by reflectance measurements. Both methods were used in this work on diamond.

In semiconductors the phonon spectrum which determines the absorption edge for indirect transitions has been resolved even at room temperature by electric field modulation of the absorption spectra. This type of experiment was suggested by Franz[4] and Keldysh[5] and is possible because the absorption coefficient depends on the electric field. An example of the technique for silicon at 300 K is given by Frova and Handler[6].

All absorption processes which involve interband transitions can produce electrical conductivity during illumination since they raise electrons to the empty conduction band. Whether or not photoconductivity is detectable will depend on many other factors which will be discussed in Chapter 5.

The interband transitions set an upper energy limit to the light which is normally used in studies of defects, but not all the other absorption bands are necessarily related to defects in the structure as both phonon and exciton absorption processes produce characteristic energy loss. A generalized set of possible optical transitions is sketched in Fig. 3.3 for an insulating material which also contains

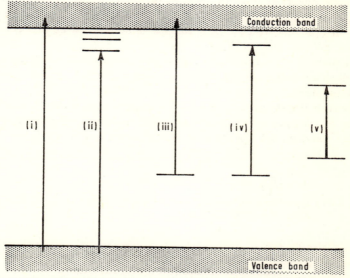

Fig. 3.3 Possible optical transitions between energy levels of an insulator containing defect levels.

defects. Some of the more obvious properties of each type of transition will now be considered.

The first case (i) is an interband transition which, as previously mentioned, is temperature sensitive and has a phonon induced tail. The absorption process produces both electrons and holes so we expect some photoconductivity. The apparent band gap determined by the increase in photocurrent with photon energy cannot be easily used to estimate the band gap because of the decrease in penetration of the light into the sample at high energies. Space charge and recombination effects (Chapter 5) which result from this mean that the photocurrent is no longer proportional to the absorption coefficient.

A number of transitions appear on the low energy edge of the band gap and these are termed exciton absorption bands. The exciton was proposed to be a bound electron-hole pair, consequently the states exist just below the conduction band; these states are labelled (ii) in Fig. 3.3. This absorption process will provide enough energy to make the pair mobile but insufficient energy to separate them. There is also some similarity between an electron bound to a positive hole and an electron bound to the proton in a hydrogen atom. The similarity extends to a series of absorption lines which converges for the higher members as $1/n^2$. In the solid the relationship is closely given by $(E_{gap} - E_{exciton}) = A/n^2$. It differs from the hydrogenic absorption in that the presence of a valence band gives wide absorption peaks, not line spectra. There is also deviation from $1/n^2$, since the excited states extend over different regions of the lattice and there is an effective change in the dielectric constant, which of course is a bulk parameter.

Exciton line resolution is possible at low temperatures, and as many as five lines in the hydrogenic series have been seen in cuprous oxide[7] and spectra appear of the form shown in Fig. 3.4.

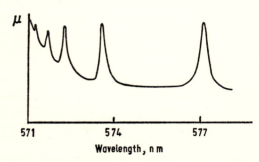

Fig. 3.4 An exciton absorption spectrum of cuprous oxide.

In many materials the onset of exciton absorption can be confused with true band to band transitions because the optical absorption coefficient rapidly rises to 100 cm^{-1}. However, the bound exciton does not produce photoconductivity. Recent developments with two photon absorption processes, described in § 3.11, can be used to separate exciton and band to band processes. The field has been reviewed both by Dexter and Knox[8] and Knox and Teegarden[9].

In the third class of absorption process the transition (iii) is from a localized level to the conduction band. We therefore expect a wide absorption band, say 1 eV wide, and associated photoconductivity at all temperatures. We also expect that the absorption of light within the band will reduce the number of electrons in the ground state. Hence the absorption process can also produce bleaching of the absorption band and luminescence when the electron decays from the conduction band.

We should be careful to distinguish between the terms 'bleaching' and 'annealing'. In both cases some treatment, heating or illumination, reduces the strength of an absorption band. If the treatment merely depopulates the ground state we term this 'bleaching'. Alternatively the treatment may irreversibly destroy the site in the lattice which showed these energy levels; in this case we designate it an annealing process.

A further property of this absorption process is that one can excite the electron by thermal means to the conduction band. Such processes as thermoluminescence and thermally stimulated currents are discussed in Chapter 4. As a guide to the temperature required to excite the process thermally we can use the rule that we need a temperature of the order of $T = E/25k$.

Returning to Fig. 3.3 we see that we can also observe transitions between levels, both of which lie in the band gap (iv). If the upper level is close to the conduction band then the transition from this to the band can be made thermally at high temperature. Therefore such pairs of levels show high-temperature photoconductivity. Predictions of the band shape and width with temperature are less simple. There will of course be a broadening of the band with increased temperatures. The absorption band shape may alter and become asymmetric with a tail towards higher photon energies if there is strong interaction between the upper level and the conduction band.

The final case of two localized levels well separated from the lattice bands (v) does not give photoconduction nor does the defect interact with other levels via the conduction band. Also any luminescence process as the excited electron returns to the ground state will give a spectrum characteristic of the defect which can only be excited by absorption of light in the one transition.

Excitation of the electron to higher levels either directly from the ground state or from the excited state can give photoconduction, luminescence etc., and such events enable us to estimate the position of the levels within the forbidden gap.

3.2 Configuration co-ordinate diagrams

A convenient way to describe the processes of optical absorption and luminescence is by a configuration co-ordinate diagram. The form of the diagram has the same general shape for many systems, but the detailed choice of variable parameter depends on the example. Such a diagram is shown in Fig. 3.5. In this the total energy of the

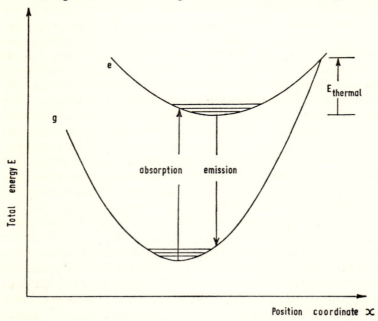

Fig. 3.5   A configuration co-ordinate diagram.

defect or entire system is plotted versus an abscissa which represents the movement of the ions or electrons from their equilibrium position. The generality of the curves is obtained because we expect parabolic energy levels around stable minima whether we consider separation of ions in a molecule or the energy levels of electrons near a defect in a solid, because for small displacements the system acts like a simple harmonic oscillator.

Before making quantitative calculations it is worth noting the more obvious features of such a diagram. In the case of electron transitions the depth to which each well is populated will depend on the temperature and density of available states. Only vertical transitions will be allowed, i.e. the Franck-Condon principle that the electronic transitions are rapid compared with any lattice relaxation. Therefore, one expects absorption and emission bands which broaden as the temperature is raised and more states are populated.

In general the weaker binding of the higher level to the centre will produce a shallower well displaced from the ground state minimum. Consequently there is a difference between the absorption and emission spectra. As the electron returns to the ground state it produces a broad, low energy emission band. This is known as a Stokes shift between the absorption and emission energies.

Despite the fact that transitions originate from discrete levels most absorption bands in insulators are several tenths of an electron volt in width, and one can confidently use semiclassical methods rather than full quantum mechanical ones to calculate the band shapes. Detailed references to the subject have been made by a number of authors[10, 11, 12, 13]. We should note that, whilst we can accurately calculate a configurational co-ordinate diagram from the absorption and emission data, we may have made no advance in understanding the nature of the centre which we are studying. An exception to this is where we begin with a lattice model and compute the wavefunctions that result. This problem was pioneered by Williams[10] in his work on thallium-doped KCl.

To calculate the absorption or emission curves, we must compute both the shape of the ground and excited states and the distribution of vibrational levels which are thermally populated. The simplest assumptions of the binding forces of the electron to a defect are Hooke's law type forces, so we have a simple harmonic oscillator of frequency v and parabolic energy bands. The ground state distribution function cannot be purely classical as this implies a sharp absorption line at low temperature as the average energy $kT/2$ approaches zero. However, the problem is resolved by considering a zero point energy to give the band a finite width of $hv/2$. The band shape is then essentially determined by the number of vibrational levels at each level. We should note that as many as 50 or 60 quanta can be involved in the process, so a random addition of the vibrational quanta produces a Gaussian-shaped distribution of filled states. Consequently the absorption band is also approximately Gaussian.

Following the model of Williams and Hebb[10] we can write an absorption probability as

$$P(E) = \left(\frac{C}{\pi kT}\right)^{\frac{1}{2}} \exp\left(-\frac{CX^2}{kT}\right)\left(\frac{dX}{dE}\right)$$

where E is an energy difference between the ground and excited state, C is a constant and X is the displacement of the electron from the equilibrium position. To adjust the classical expression to match the quantum mechanical expression we use an effective temperature

$$T_1 = \frac{h\nu}{2k} \coth\left(\frac{h\nu}{2kT}\right)$$

where ν is the vibrational oscillator frequency. This now allows us to predict the shift in peak energy of the absorption with temperature, also the half width of the absorption band is proportional to

$$\left[\coth\left(\frac{h\nu}{2kT}\right)\right]^{\frac{1}{2}}.$$

A selection of the absorption curves calculated by Williams and Hebb[10] for KCl : Tl are shown in Fig. 3.6. The difference between

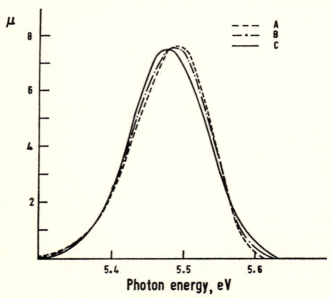

Fig. 3.6   Calculated absorption band shapes for KCl : Tl.
  Curve (A) is for a semiclassical parabolic potential,
       (B) a semiclassical linear potential,
       (C) a quantum mechanical treatment.

a classical calculation with a parabolic upper state and a tangent to the parabola is trivial, and even the more detailed quantum mechanical treatment produces a similar absorption band.

A more advanced analysis of the vibrational spectrum is possible in principle by accurate determinations of the band shape which should reveal features of the spin-orbit coupling of the excited state to the lattice. A description of this method, using the moments of the band, is given in a review by Henry and Slichter[14].

3.3 Oscillator directions and polarized absorption; dichroism

Electronic transitions which occur within the constraints of a solid will show the symmetry of the defect site within the solid. This is readily shown by measurements with polarized light which stimulate only those electric dipole transitions which have axes oriented along the electric vector of the light. However, the dipole moment need not have the same symmetry axis as the defect and positive identification of the defect symmetry is not trivial. In the well-known studies of the M centre[15] in alkali halides several groups used polarized absorption and emission measurements to justify alternative models for the centre. Finally one model was substantiated by measurements

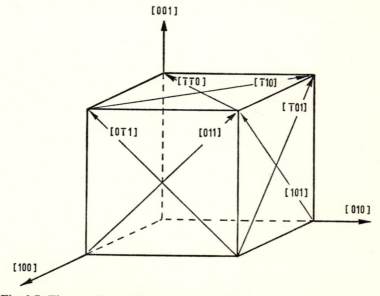

Fig. 3.7 The possible oscillator directions of $<110>$ dipoles in a cubic crystal.

on the electric dipole moment of the centre in the ground state, and by ESR and ENDOR experiments on the centre in the excited state. If the light can either bleach the centre or allow it to reorient in one of the equivalent directions, then prolonged irradiation can produce a change in the populations of the equivalent sites. This will be shown by subsequent measurements with light polarized in other directions and is termed dichroism.

For illustration we can consider a cubic crystal with defects which have axes in <110> directions. There are six such directions which absorb light at rates given by $\cos^2 \theta$ where θ is the angle between the oscillator and the polarized light.

By reference to Fig. 3.7 we see that light along the [100] direction does not interact equally with all six types of oscillator. If it is polarized along [001] then the <011>, <0$\bar{1}$1>, <$\bar{1}$01> and <101> all absorb equal amounts with a component intensity reduced by $\cos^2$ 45 $(=\frac{1}{2})$. A subsequent probing measurement of the dichroism with light along [001] and [010] shows less absorption of [001] light because the <$\bar{1}\bar{1}$0> and <$\bar{1}$10> centres are only 'seen' by the [010] light and were not bleached. Initially the bleaching of the [001] absorption is twice as fast as the [010] absorption. Eventually all the [001] absorption bands would bleach and the [010] level would be reduced to half the initial value.

For a defect with an axis which we suspect is <110> we would also try bleaching with [011] type light. The dichroism viewed with [011] and [0$\bar{1}$1] light would give an initial rate of [011] bleaching five times that of [0$\bar{1}$1] bleaching with finally no [011] absorption and a halved [0$\bar{1}$1] absorption. Unless one measures the entire bleaching curve it is not possible to make a positive identification of the axes. For example it is shown in the review paper by Compton and Rabin[15] that one cannot necessarily separate a <110> axis from a ⟨121⟩ axis without a very full set of measurements.

Similar arguments of the relationship between polarized absorption and polarized emission from a defect may be useful, but experience with the M centre indicates that interpretation is not unequivocal.

In non-cubic materials there is a greater chance of success because the problems of bleaching, reorientation or overlapping absorption bands are reduced[16]. As before we can choose models for the defect and compute the absorption by electrons directed along particular bond directions. However, with lower crystal symmetry there is anisotropic absorption before bleaching. For example in ruby the chromium occupies a substitutional position in the Al_2O_3 lattice. The host lattice has a rhombohedral structure and measurements with light polarized parallel or perpendicular to the c axis show the anisotropic absorption spectra of Fig. 3.8. The emission spectrum is also polarized[17].

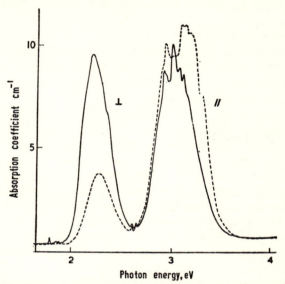

Fig. 3.8 The absorption spectrum of ruby measured with light with the electric vector polarized either parallel or perpendicular to the c axis of the crystal.

3.4 Smakula's equation and the f number

The theoretical shape of the optical absorption band and the result-ant strength of the absorption can be computed moderately well, and estimates of the defect concentration from the measured absorption curves are generally made by comparison with the equations de-veloped by Smakula. The problem is to calculate the transition probability between the range of ground and excited states described by the configurational co-ordinate diagram. That is, to evaluate the transition probability per second.

$$\underset{g \to e}{P} \propto \left| \int \psi^*_{\text{excited}} H_{\text{interaction}} \psi_{\text{ground}} \, d\tau \right|^2$$

$\times$ (density of states at the transition energy).

From this we can write the absorption cross section per incident photon in unit volume of material as $\sigma_{g \to e} \propto P_{g \to e} \times h\nu$.

Thus the total absorption across the absorption band is

$$\sigma_{g \to e} \propto h\nu \, |r_{ge}|^2 F_{ge}(h\nu)$$

where $F_{ge}(h\nu)$ is the shape function of the absorption line and r_{ge} is the position co-ordinate for this part of the configuration co-ordinate diagram.

As a result of absorbing the light energy the lattice becomes polarized. In the quantum mechanical expression for the polarizability this is related to the dipole moment for the transitions by[12]

$$\alpha = \sum \frac{2e^2 |r_{ge}|^2}{3(E_e - E_g)}.$$

However, from a classical model we could have derived the low frequency polarizability as

$$\alpha = \sum \frac{e^2 f_{ge}}{m\omega^2_{ge}},$$

where f_{ge} refers to the number of oscillators at each of the allowed frequencies. From these two expressions we then write an 'oscillator strength f' for the overall transitions over the entire range of energies,

where

$$f = \frac{4m\pi}{3h} \omega_{ge} |r_{ge}|^2.$$

For a single atom with one electron involved in absorption processes $\sum_e f_{ge} = 1$ if several higher states are allowed.

If the oscillator is embedded in a solid then the effective mass of the electron, the dielectric constant, and the refractive index n of the solid modify the absorption cross section. With these additions we can compute the strength of the absorption band for N centres of effective oscillator strength f. If the absorption coefficient is μ at the photon energy E, then the expression derived by Dexter[12] is

$$Nf \frac{(n^2 + 2)^2}{9} \frac{2}{n} \frac{\pi^2 e^2 \hbar}{cm^*} = \int \mu\,(E)\,\mathrm{d}E.$$

Evaluating the constants in centres per cubic centimetre this gives

$$Nf = 8 \cdot 21 \times 10^{16} \frac{n}{(n^2 + 2)^2} \int \mu\,(E)\,\mathrm{d}E.$$

For simplicity we replace the integral by an equivalent area of (constant) × (maximum absorption coefficient) × (width at half maximum). That is to say the effective number of defects = const $\times \mu_{max} \times W$. This is known as Smakula's equation and takes the form

$$Nf = 1.29 \times 10^{17} \frac{n}{(n^2 + 2)^2} \mu_{max} W \quad \text{for a Lorentzian band,}$$

$$Nf = 0.87 \times 10^{17} \frac{n}{(n^2 + 2)^2} \mu_{max} W \quad \text{for a Gaussian band.}$$

Measured absorption bands tend to be between these two ideal curve shapes.

3.5 Absorption coefficient and the defect concentration

From Smakula's equation it seems possible that we can estimate the defect concentration in the solid. Absorption band shapes can generally be resolved even with overlapping bands, so the only remaining problem is to determine the oscillator strength. This is a difficult and major problem which can be insoluble, with the result that the absorption coefficient merely indicates the relative concentration of a particular defect.

The most direct approach to finding the f value is to look at absorption bands which result from impurities. Ideally, control of the dopant gives a known concentration of impurity ions in the solid, but the limitations to this are

(i) only a fraction of the ions occupy the correct site in the lattice.
(ii) not all the ions exist in the correct charge state.
(iii) ions may form colloids or associate with other defects.

Such problems are fairly obvious, and it is not difficult to speculate on the range of sites available in the lattice which will accommodate impurity ions. The factors of atomic size and valence are helpful in these guesses. However, it is more difficult to predict when agglomeration, complex formation or phase precipitation will occur. These same considerations indicate that impurity ions may preferentially form complexes with lattice vacancies in order to minimize the distortion of the lattice. Examples of such centres are the Z bands in alkali halides, which are vacancy centres linked to impurities. These will be mentioned in Chapter 7. Saturation or colloid formation may be apparent at very low dopant levels, for example in lithium fluoride Johnston[18] found that magnesium doping to 100 parts per million was possible, but at higher impurity concentrations he observed colloidal metal aggregates. It should be apparent that the nucleation rate of the impurities depends on the number of available sites, so that the rate of quenching the melt can change the equilibrium level. In single crystals the dislocation density and grain size will be important. There are thus no clear rules to predict the saturation level.

Alternatively we can look at results of impurity doping where the ionic size and valency of the extra ions can be readily accommodated by substitution into the lattice. One such example is ruby. Ruby is Al_2O_3 which contains chromium as a substitutional ion on aluminium lattice sites. A small lattice strain is introduced by replacing the aluminium by a chromium ion, but up to 1% chromium can be accommodated. The system is particularly interesting because the chromium ion has an incomplete inner shell and so gives both line absorption and line emission spectra which are modified by the surrounding lattice. We find that the line width increases with increasing chromium concentration, in addition new fluorescence lines appear (at 700·4 and 703·3 nm) at high chromium concentrations. These increase in intensity at a faster rate than the simple red lines (termed R_1 and R_2) as one increases the dopant level. Schawlow[19] interpreted the new lines as the result of an exchange coupling between adjacent pairs of chromium ions and considered that the enhanced appearance of such lines at higher concentrations suggests that the pairing of the chromium is not a random process but is a preferred process for the inclusion of the chromium. Such effects produce a change in the oscillator strength of the absorption bands.

A large change in oscillator value with defect concentration has been reported[20] for KCl : Tl. Here the interaction between adjacent thallium centres produces new optical absorption bands. In this case we are viewing a normally forbidden transition, so apparent changes in oscillator strength may result from either a change in the selection rules or the production of new absorption bands when the thallium centres interact. The numbers quoted for two absorption bands at 196 nm and 250 nm are

$$f_{196 \text{ nm}} \approx 1 \qquad f_{250 \text{ nm}} \approx 0\cdot1 \text{ for } 2 \times 10^{-4} \text{ Tl}^+ \text{ per cm}^3.$$
$$f_{196 \text{ nm}} \approx 0\cdot3 \qquad f_{250 \text{ nm}} \approx 0\cdot07 \text{ for } 10^{-2} \text{ Tl}^+ \text{ per cm}^3.$$

There is a further problem that the intensity of the absorption band gives a measure of the number of defects in a particular charge state and the excitation conditions of temperature, excitation energy and flux determine the final equilibrium between the possible states. The obvious example of this is the halogen vacancy in the alkali halides. This can exist as any of the following centres:

α centre	a halogen vacancy
F centre	a halogen vacancy with a trapped electron
F' centre	a halogen vacancy with two trapped electrons
M centre	two adjacent F centres
R centres	three adjacent F centres
N centres	four adjacent F centres

Such interaction between defects often implies a close spatial association and a consequent modification of the f value. However, it also allows us to proceed from a 'known' defect concentration (e.g. F centres produced by chemical doping) to a quantitative estimate of the N and f for a new absorption band, in this example the conversion of F centres to M centres, etc.

A final minor note of caution in quantitative estimates of defect concentration is that with light polarized in one plane we only detect oscillator directions in that plane. Measurements with polarized light or in crystals of non-cubic symmetry must be made in several crystallographic directions.

3.6 Measurement of the absorption coefficient

We have spoken in terms of an absorption coefficient whereas in practice we measure the fraction of the light transmitted by a crystal at a particular wavelength. To relate these two quantities consider the fraction of the energy absorbed per centimetre as μ, so that for a thickness dx the change in light intensity is

$$-dI = \mu I_x dx$$

or integrating within the limits of the sample, as in Fig. 3.9,

$$I_t' = I_0' e^{-\mu t}.$$

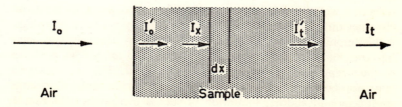

Fig. 3.9 The passage of a light beam through an absorbing medium.

This is known as Beer's law of absorption. However, a fraction R of the light is reflected at each interface, so that the total transmitted light is

$$I_t = I_0(1 - R)^2 e^{-\mu t} + I_0(1 - R)^2 e^{-\mu t} R^2 e^{-2\mu t} + \cdots$$

This is a geometric progression and sums to

$$I_t = \frac{(1 - R)^2 e^{-\mu t}}{1 - R^2 e^{-2\mu t}} I_0.$$

For low reflectivity materials, such as alkali halides, $R \approx 5\%$, so the approximation

$$I_t = I_0 e^{-\mu t}$$

is adequate.

For high reflectivity material such as rutile, $R \approx 20\%$, and then the full expression must be used for absorption measurements.

As an indication of the magnitude of the absorption coefficient which is measured we can expect $\mu \sim 10^6 \text{ cm}^{-1}$ for band to band transitions, $\mu \sim 10^4 \text{ cm}^{-1}$ for exciton bands and $\mu \sim 1$ to 100 cm^{-1} for normal absorption bands. Commercial spectrophotometers which measure the transmitted light as a function of wavelength are normally reliable to transmissions of less than 1%, and the best can be stretched to $10^{-4}\%$ with careful treatment.

As a first order correction to the reflection coefficient it is common practice to use a dummy sample in the reference beam of the spectrometer. Many instruments also present the results directly on a logarithmic scale on the assumption that Beer's law is adequate. In these cases the measured quantity is called an optical density or absorbance and written as

$$D = \log\left(\frac{I_0}{I_t}\right) = \frac{\mu t}{2 \cdot 3}.$$

The above expressions were developed for light normally incident on plane parallel specimens. For rough material it is possible to use an integrating sphere[21]. An integrating sphere should have totally reflecting walls so that when we shine in monochromatic light the chamber diffuses the light to reach a uniform power density which is then monitored by a photodetector. If we now insert the specimen into the chamber, the power level is reduced and this gives a measure of the strength of the optical absorption band. The measurement may not be quantitative for very rough and strongly absorbing specimens and another limitation of such a measurement is that a false value of the absorption is obtained if the excited state decays

by a luminescence process. For example, if the minima of the ground
and excited states lie vertically above one another then the photon
density within the sphere is unaltered after a time equivalent to the
excited state lifetime (typically 10^{-8} s.).

Similarly false readings occur if any of the luminescence is detected
and the simplest solution to this problem is to analyse the emitted
light with a second monochromator. The alternative deconvolution
of the simple signal requires that we know the emission spectrum and
the wavelength response of the detection system.

3.7 The growth of colour centres with irradiation

In this section we shall consider the intensity of absorption bands
as a function of incident radiation and show that it is possible to
distinguish between defects formed by irradiation and existing defects which are merely made visible by charge capture. Particular
models of defect formation will not be considered until Chapter 8.
As was mentioned above, the charge state or association with other

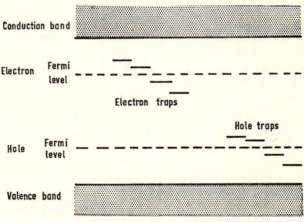

Fig. 3.10 Trapping levels and the Fermi levels for defect sites in the
         crystal.

defects is a complex function of the number of defects, impurities,
dislocations, temperature and irradiation conditions.

It is clear from an energy band picture of an insulator that the
impurity levels occur at all points in the forbidden energy gap.
However, depending on the position of the Fermi levels, only certain
traps will actually contain any charge, as is indicated in Fig. 3.10.
Any form of energy which redistributes the charges in the crystal also

alters the Fermi levels. If the irradiation stimulates electrons and holes into the lattice bands, then, firstly, the Fermi level moves and, secondly, the empty traps capture charge at a rate proportional to their respective capture cross sections. Many sites will capture both electrons and holes and so form recombination centres. Here we must consider the concentrations and cross sections of both types of charge which result from the total defect distribution in the crystal and the mobility of the charges. The problem is conceptually simple, but in practice it is impossible to calculate the absolute changes in population that occur in real crystals.

Not only do we rarely know the impurity content of a sample to better than a few parts per million, but also we only have approximate ideas of the distribution of the defects before irradiation. The fact that ionic crystals have dislocation lines which are charged means that there is a vacancy or impurity region around the dislocation to compensate for the space charge. Calculations by Kliewer[22] show that both the sign and magnitude of these fields depends on the type of impurity and the temperature. One direct consequence of these fields is that electrons and holes migrate in opposite directions and even traps which exist in small concentrations may capture charges before recombination can occur. For small crystallites of silver halide these electric fields are thought to exist from the surface into the centre of the crystallites, and separate the holes to surface traps and preferentially direct the electrons to special sites at which a latent image can form. If we consider the photographic process as the liberation of an electron-hole pair by the incoming photon we expect the hole to go to the surface and the electron to be trapped at a silver ion, making a neutral atom $(Ag^+ + e^- \rightarrow Ag^0)$. The silver atom may be mobile and unstable, but will stabilize once it associates with three or four other silver atoms to make a colloidal silver speck (i.e. the latent image). This must happen in a time short compared with the lifetime of a free silver atom, or atoms. So a latent image can only form if the light intensity is sufficient to produce at least four electrons in the grain within this lifetime. At medium light intensities the film will blacken at a linear rate with dosage, but at very high light levels the linearity is destroyed because the large hole concentrations produce recombination and cluster breakdown so the crystallites actually form fewer latent images. This problem is known as solarization. No image appears even for prolonged irradiation if the flux is too small, and at low light levels the problem is known as reciprocity failure.

Control of the hole trapping levels offers a means of controlling film speed, and emulsions become more sensitive if one adds sulphur, etc., to the silver bromide. The sulphur is effective even though it sits on the surface of the grain because of the internal electric field. The efficiency of latent image formation versus light intensity is shown in Fig. 3.11.

Flux-dependent effects are thus to be expected if there is competition between various defects for the free charges. Once again the alkali halides provide a well-documented[15] example of the phenomenon for the series of halogen vacancy centres F, M, R, N (i.e.

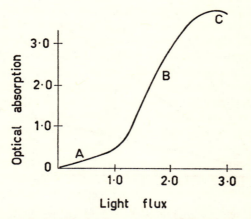

Fig. 3.11   The efficiency of forming a photographic image as a function light intensity. In the region A only a few grains develop, region B is linear with dosage and solarization occurs at region C.

single vacancy with a trapped electron, a pair, triplet and quartet of F centres). Interconversion of defects in the series is possible by bleaching one centre via its optical absorption band, and although the system is reversible the rate constant for bleaching the larger units is less than the small ones, so that the crystal tends to contain the associated centres after a series of bleachings.

From Sonder and Sibley's results[23] of Fig. 3.12, we conclude that an irradiation treatment produces a certain number of defects, but the observable fraction is flux-dependent. Under all conditions there is a quadratic relationship between the M and F centre growth curves, but the slope of the line depends on the rate of ionization as is clear from the lines E_1, E_2 and G. Sonder and Sibley demonstrated that the equilibrium balance between the centres was flux-dependent by the following experiment. A crystal was irradiated with 1·5 MeV electrons at a flux of 0·34 μA cm^{-2} to a point P on the growth curve E_2 and then electron irradiated for a further 25 s at 0·06 μA cm^{-2}. This dose was trivial compared with the total dosage, but a major change in M and F centre populations occurred. Also the new equili-

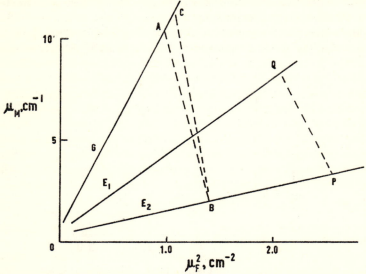

Fig. 3.12 The relative growth of F and M centres in KCl during G, gamma ray irradiation of 4·5 × 10^6 R per hour, E_1, E_2, 1·5 MeV electron irradiation at fluxes of 0·06 μA cm^{-2} and 0·34 μA cm^{-2}. Transitions A, B, C and PQ are described in the text.

brium point Q corresponded to the expected point computed for the total dose having been given at the lower flux (i.e. growth curve E_1). The procedure could be reversed and the original coloration point P was reached by a short irradiation at the higher flux. A similar change in equilibrium can be established between electron and gamma ray growth curves as is indicated by the sequence ABC.

Thommen[24] noticed that the balance could also be reversibly altered by changes in the temperature of the sample with a constant irradiation flux.

Similar problems exist in absolute intensity measurements of luminescence, although in this case we might more readily expect that both the fraction of radiative decay processes and the number of trapping levels would vary with irradiation flux and sample impurity.

The equilibrium state may also be influenced by defect motion, if the defects have different mobilities in the various charge states. Two examples of this are the interstitial silver in silver bromide, which is mobile as a neutral atom, and the halogen vacancy in alkali halides, which is trapped when it contains an electron but is very mobile (at 300 K) as an empty vacancy centre.

Excluding flux-dependent effects we may now predict how an absorption band will develop as a function of irradiation dose. For intrinsic defects such as vacancies there may already be a number of sites in the crystal which were a consequence of cooling the crystal from the melt. This concentration is readily calculated from a thermo-dynamic knowledge of the formation energy of the defect and the subsequent heat treatment of the sample. Irradiation of quite low energy which releases free charges will allow this defect to produce an absorption band which saturates when the fixed number of sites comes to equilibrium with the other trapping processes[25]. This means a growth curve for the absorption constant μ with time as

$$\mu = \mu_{max}(1 - e^{-\alpha t}).$$

The constant α will depend on the competitive process, so it can be altered by the impurity concentration.

Similarly there will be a saturation concentration of defects which arises from the dissociation of complexes or defects generated or released from dislocation sites. These yield a second term

$$\mu = \mu'_{max}(1 - e^{-\beta t}).$$

Finally, new defects may be produced by the irradiation from the bulk material at a rate given by

$$\mu = \gamma t.$$

In practice even prolonged irradiation may not lead us above the saturation levels of μ_{max} and μ'_{max} to the linear region of newly produced defects. Because, if the total defect concentration is very large, then there will be interaction between defects and 'radiation annealing', e.g. new vacancies absorbing existing interstitial atoms. Such effects are usually apparent by defect concentrations of 0.1% of the lattice sites.

Below this maximum defect limit the various terms can produce a variety of colouring curves such as those illustrated in Fig. 3.13(a). This figure shows the two types of coloration curve where curve A

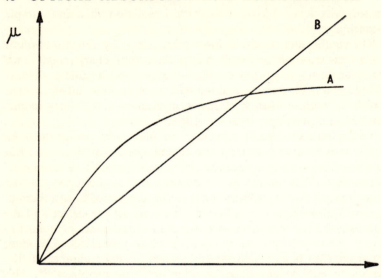

Fig. 3.13(a) The growth curves for optical absorption bands which, A, saturate because of trap filling and, B, are generated by the irradiation.

is saturable and B is for linear defect production with irradiation dose. The rate constants α, β, γ and the saturation values of the components μ_{max} and μ'_{max} are specimen-dependent, so that the

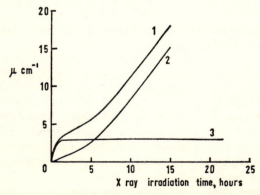

Fig. 3.13(b) An example of a growth curve for a KCl crystal irradiated with X-rays. The original curve 1 has been analysed into two components 2 and 3.

resulting envelope of curves may have a variety of shapes. One generally assumes that the linear defect production rate γ is independent of the other stages, and changes in μ_{max}, μ'_{max}, α, β and γ can be detected by experiments where doping or straining of the sample is possible. An example of the analysis of a coloration curve is shown in Fig. 3.13(b) for the growth of the F band in KCl during X-ray irradiation. The measured coloration was curve 1. This was interpreted as a saturable portion, curve 3, and a 'linear' part in curve 2. As a first estimate this is adequate, since the fast colouring first stage is so different in rate from the slow colouring second stage, but there may be some doubt about the size or number of saturable components. This linear part represents the rate of new defect formation γ. Experimentally it may be determined, as in this example, by prolonged irradiation. An alternative approach is to reduce the crystal temperature so that the defects are immobile and do not cluster or migrate to dislocations. In some instances this suppresses part of the saturable components.

Pooley[26] has considered the influence of impurities on the growth rate of the F band in alkali halides from the effect of competitive charge capture by the halogen vacancy or impurity. In his model different alkali halides show different sensitivity to impurities, for example in KCl the linear rate is only altered for impurity concentrations above 0·25 p.p.m., whereas in KI effects are detectable at 0·006 p.p.m. Such calculations demonstrate that 'chemically' pure crystals are inadequate for many colour centre studies.

3.8 Simple calculations of energy levels

Possibly the simplest defect that we can consider is an electron trapped in a vacancy. As a first approximation we shall assume the system behaves like an electron in a box, i.e. a simple square well potential. The familiar solution of the Schrödinger equation for this example gives the energy levels of the system as

$$E = \frac{h^2}{8mL^2}(n_1{}^2 + n_2{}^2 + n_3{}^2),$$

where h is Planck's constant, m the electron mass, L the dimensions of the box and n_1, n_2, n_3 are the quantum numbers. The first possible transitions will be from the 111 state to the degenerate set of levels, 112, 211 or 121.

To decide if this approximation is reasonable, we can use the known energy of the first F centre transition for KCl and decide what is the size of the box that we require. If the F band occurs at 2·3 eV we calculate L as 0·7 nm. This is sensible since the electron is not entirely confined within the first set of neighbours (see Chapter 2), and the lattice spacing of KCl is 0·63 nm. Even this simple estimate of the energy levels is useful, since we can now predict the position of the absorption band within 20%. An extension of the square well model to M centres, a pair of F centres, also shows we expect the absorption band at roughly half the energy of the F band.

A very valuable prediction of the model is that E is proportional to L^{-2}, so that in a series of compounds with the same lattice structure there is a steady movement of the energy of the peak position from one compound to the next, which is simply determined by the lattice spacing. This greatly reduces our problems of defect identification, for we only need to make a definitive study of a defect in one compound and the same defect will have been identified in the remainder of the series.

A great deal of effort has been made in the theoretical calculations to improve on the accuracy of the peak energy and the shape of the absorption band of the various colour centres. However, for our purposes the reduction of the 20% error in this simple calculation

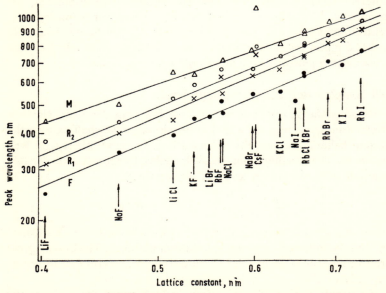

Fig. 3.14 The Mollwo-Ivey law for F, R_1, R_2 and M centres in the f.c.c. alkali halides.

is unimportant and for more detailed discussion the reader is referred to the review article by Markham[27], and more recent papers by Fowler[28], and Parker and Dawber[29].

3.9 Mollwo-Ivey laws

The simple relationship between absorption energy and lattice parameter ($E \propto L^{-2}$) outlined in the previous section has been detected for many types of imperfection. Once again the literature exists mainly for alkali halides and empirical results show that the inverse square law is still reasonable. The law was first noted by Mollwo[30] and Ivey[31] and so bears their name. Examples of the peak dependence of the alkali halide F and M centres with lattice spacing are shown in Fig. 3.14. Only those with a face-centred cubic structure are considered. Typical exponents for the slopes are $-1\cdot84$, $-1\cdot51$, $-1\cdot76$ and $-1\cdot61$, for the F, M and R_1 and R_2 centres. The theoretical justification for this scaling is only possible, if the potential is a simple function of r/d, where r is the distance

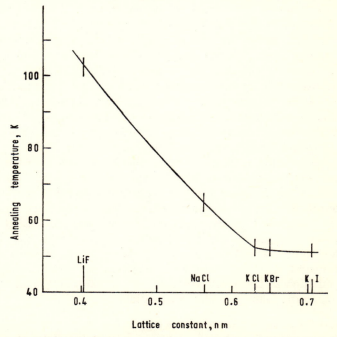

Fig. 3.15 Some annealing temperatures quoted for removal of H centres in alkali halides.

co-ordinate and d is the lattice constant. For large cations correction terms are necessary which override this simple empirical expression. The extended theory given by Bartram et al.[32] can cope with the case of the alkaline earth fluorides and oxides where the simple law led to an incorrect determination of the F band (see Chapter 7).

It is interesting to note that some continuity of features from one alkali halide to the next has also been reported[33] for the defect annealing stages of the alkali halide defects; for example, in Fig. 3.15 is shown the temperature at which H centres have been said to anneal. There is a general trend to higher annealing energies for the smaller lattice parameters, which is also found for some of the F' and V_K centres, but rarely do the centres anneal in a single stage, and so there are severe doubts as to whether the stages quoted represent 'pure' annealing or annealing of centres linked to impurities. These effects are discussed in more detail in § 7.3 for V_K centres.

3.10 Excited state spectroscopy

Normal optical absorption bands result from the transfer of energy from a single photon to an electron in a ground state energy level at some lattice site. Consequently the absorption data give information on the ground state. It is clear that we could make similar spectroscopic studies (i.e. optical, ESR, ENDOR) of the excited state by observing transitions which are stimulated to higher levels. The only experimental problem has been that we need a large concentration of defects in the excited state. This can be achieved if the excited state lifetime is very long or we have a very intense source of light to stimulate electrons from the ground state.

The stimulation is simple if the absorption band overlaps the emission line of an intense light source, such as a laser. All frequencies within the band will promote electrons to the excited state, but the efficiency is greatest at the high absorption coefficients at the centre of the band. Laser pulses of 10^{20} photons are feasible for many wavelengths, so an entire defect concentration (say 10^{16} centres) can be excited. A simultaneous measurement of absorption spectra will then give data on the excited state and its environment.

In addition to the usual symmetry and coupling information we can obtain the following extra information:

  (i)   ESR of two electron centres;
 (ii)   access to energy levels masked by band to band transitions;
(iii)   separation of overlapping absorption bands;

    (iv) identification of absorption bands arising from the same ground state;

    (v) transparency in normally absorbing material;

    (vi) emission studies from higher levels.

Some expansion of these topics will now be given.

    (i) As was emphasized in Chapter 2 the techniques of ESR and ENDOR provide an unequivocal model for a defect, but it must have at least one unpaired electron. In the ground state of a two-electron centre the spins are paired and ESR is impossible. However, the triplet excited state removes this degeneracy and ESR and ENDOR are possible (some care is still necessary to ensure that the ground and excited states refer to a simple defect, i.e. not a mixture of states from a close pair of defects). Successful resonance experiments were used[34] to establish the models of the alkali halide M and R centres as a pair and triplet of F centres.

    (ii) From Fig. 3.16 we see that a sequential absorption of two

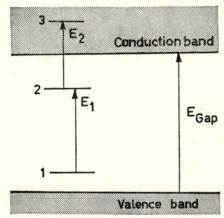

Fig. 3.16 A two-stage absorption process.

photons of energy E_1 and E_2 can lead to the population of an upper state, labelled 3, which is hidden in a conduction band, Both photon energies E_1 and E_2 are less than the band gap E_g, so the only absorption processes take place at the defect. Direct access to level three from the ground state with a photon of energy $(E_1 + E_2)$ would not be possible, since this energy is greater than the band gap and most photons would be absorbed at normal lattice sites and promote band to band transitions. The two-photon study of level three avoids this problem.

(iii) At high photon fluxes (or long excited state lifetimes) complete population inversion between the ground and excited states of a defect is possible. During this situation the absorption band is reduced to zero, and simultaneous measurements of other absorption bands can then be made without band overlap such as that shown in Fig. 3.17. A normal one-photon absorption spectrum might provide

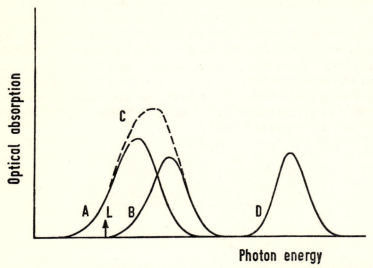

Fig. 3.17 Optical absorption bands which may be resolved by illumination of the point L with intense light; as described in the text.

curve C. From this we could not distinguish between an asymmetric curve C or an overlapping pair of bands A and B. However, C could be bleached by intense light at any wavelength in the band, but in the case of two bands B would remain if the bleaching were made at the point indicated by L.

(iv) Similarly we can remove all absorption bands which have a common ground state but different upper states. For example, if both bands A and D of Fig. 3.17 have a common ground state then during the bleaching pulse only band B will remain. We should also note that any wavelength in either band is suitable for bleaching both bands.

(v) As we have seen, an intense laser pulse can bleach all absorption bands at one wavelength so that a formerly opaque material will become transparent. An interesting application of this effect is in a fibre optic communications system. The object here is to utilize the wide bandwidth of a laser beam which acts as a carrier wave. Unfortunately the absorption bands which occur in the glass fibres may attenuate the signal by unacceptable amounts. It seems that such a problem could be avoided by a strong pulse which precedes the signal pulse. This would remove the absorption bands and make the system transparent.

This solid state example is possible because the light will not stimulate emission at the same frequency, since for defects in solids there is a large Stokes shift between the position of the absorption and emission spectra.

(vi) Finally, emission studies may be made from an upper state (say state 3 of Fig. 3.16) to all the possible intermediate or ground states of the excited system. By this process we may produce luminescence with photon energies greatly in excess of the individual energies required to stimulate the excitation. In principle the luminescence energy could exceed the band gap, but it would not escape from the bulk of the crystal. However, this technique could produce a solid state source of light in the vacuum ultra-violet. Research into this spectral region has always been hindered by inconvenient or low power light sources and, for example, complex experiments which need both ultra high vacuum and high energy photons may have conflicting design requirements. Development of solid state sources operating by excited state spectroscopy may ease this problem.

3.11 Two-photon absorption

Thus far we have only discussed conventional absorption processes involving two independent photons. However, at the high fluxes obtainable in laser beams there is a non-linear optical effect of simultaneous two-photon absorption. In this the virtual state between the ground and upper level need not correspond to a normal excited state. It is important to realize that for electric dipole transitions produced with one photon the ground and excited states have different parity, but in a two-photon absorption the states have the same parity. Complementary studies of the same energy range with both one- and two-photon absorption allows us to separate two types of energy level. During the last decade this has been used mainly to study the intrinsic exciton absorption bands in alkali halides.

In principle the experiment is to use photons from an intense laser beam and add to them secondary photons from a variable wavelength source. As we scan the spectral range with the second wavelength the intensity of the transmitted light (monitored by this wavelength) gives the absorption spectra of the crystal in the energy range of the sum of the two energies. The limitation is that the two primary sources are not themselves absorbed by the crystal. For pure materials only intrinsic processes are possible, so one can sum the two-photon energies to probe to levels at several electron volts above the conduction band edge. More sophisticated interpretation of the band structure is possible if we use polarized light.

To date the range of experiments is limited by available laser sources, but these already extend to 3·41 eV, and future developments and more efficient second harmonic generators could increase this considerably.

As an example of the technique we see from Fig. 3.18 the possible

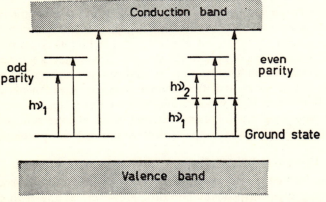

Fig. 3.18 Examples of one- and two-photon exciton transitions. Only odd parity states are reached with a single photon, but different even parity states are accessible on adding two photons.

exciton transitions viewable by one- and two-photon processes. The exciton transitions are clearly separated from the band to band absorption in the experiments of Hopfield and Worlock[35] for CsI shown in Fig. 3.19.

This type of absorption spectroscopy is still experimentally complex, but will undoubtedly receive much more attention in the future. A detailed introduction to the subject is given in the very readable review article by Mahr[36].

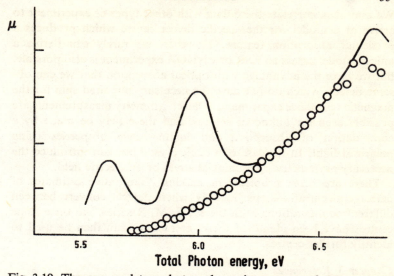

Fig. 3.19 The one- and two-photon absorption spectra of CsI at room temperature. The single photon process is shown by a continuous line.

3.12 Studies of transient colour centres

At this stage it will be evident that many defects are unstable entities and rapidly change to new equilibrium charge states or rearrange in the lattice. For a full understanding of the processes involved it is necessary to study many defects at the moment of formation. For example on irradiation of an alkali halide we expect to form the two simplest centres of a halogen vacancy (α or F centre) and a halogen interstitial centre (the H centre). The α and H are highly unstable at normal temperatures, so most measurements reveal some later and more complex stage of development. Refined experimental techniques of pulsed irradiation and measurement have been developed to cope with this situation, and these type of experiments are described by Ueta[37, 38].

3.13 Summary

Optical absorption measurements are relatively simple to make, but none the less they give considerable information on the symmetry, concentrations and interactions of the various centres and because the transition energies produce optical absorption bands in different spectral regions the properties of different defects may be resolved.

We may then correlate these data with other types of experiment to arrive at a model for the specific defect centre which produces a particular absorption feature. However, we rarely can derive a unique model unless an ESR or ENDOR experiment is also possible, but we have the advantage with optical absorption that we can observe defects which do not have the necessary unpaired spin for the magnetic resonance experiments. Defect symmetry measurements are possible even for unknown centres, and these may be made by a combination of absorption and luminescence properties using polarized light. In the case of a cubic crystal one can introduce the necessary asymmetry by a uniaxial stress or an electric field.

There are some problems in making quantitative estimates of defect concentrations, particularly when defects convert between different configurations as in the F', F, M, R series, but once some centres have been identified these interactions can then be used to identify further centres.

Chapter 3 References

[1] Bube, R. H., *Photoconductivity of Solids*, Chapter 7 (Wiley), 1960.

[2] Hopfield, J. J., *Comments on Sol. Stat. Phys.* **1**, 16, 1968.

[3] Clark, C. D., Dean, P. J. and Harris, P. V., *Proc. Phys. Soc.* **A277**, 312, 1964.

[4] Franz, W., *Z. Naturforsch.* **13a**, 484, 1958.

[5] Keldysh, L. V., *Sov. Phys.* JETP **7**, 788, 1958.

[6] Frova, A. and Handler, P., *Phys. Rev. Letters* **14**, 178, 1965.

[7] Nikitine, S., *Prog. in Semiconductors* **6**, 233, 1962.

[8] Dexter, D. L. and Knox, R. S., *Excitons* (Wiley), 1965.

[9] Fowler, W. B., Editor, *Physics of Color Centers* (Academic Press) 1968. Knox, R. S. and Teegarden, K. J., Chapter 1.

[10] Williams, F. E. and Hebb, M. H., *Phys. Rev.* **84**, 1181, 1951.

[11] Klick, C. C. and Schulman, J. H., *S.S.P.*, **5**, 97, 1957.

[12] Dexter, D. L., *S.S.P.*, **6**, 353, 1958.

[13] Curie, D., *Luminescence in Crystals* (Methuen) 1963.

[14] Fowler, W. B., Editor, *Physics of Color Centers* (Academic Press) 1968. Henry, C. H. and Slichter, C. P., Chapter 6.

[15] Compton, W. D. and Rabin, H., *S.S.P.* **16**, 121, 1964.

[16] Mitchell, E. W. J., Rigden J. D. and Townsend, P. D., *Phil. Mag.* **5**, 1013, 1960.

[17] Sugano, S. and Tsujikawa, I., *J. Phys. Soc. Japan* **13**, 899, 1958.

[18] Johnston, W. G., *J. Appl. Phys.* **33**, 2050, 1962.

[19] Schawlow, A. L., Wood D. L. and Clogston, A. M., *Phys. Rev. Letters* **3**, 271, 1959.

[20] Pringsheim, P. and Joshi, R. V., cited by Curie D., *Luminescence in Crystals*, p. 70. (Methuen) 1963.

[21] Bastin, J. A., Mitchell, E. W. J. and Whitehouse, J. E., *Brit. J. Appl. Phys.* **10**, 412, 1959.

[22] Kliewer, K. L., *J. Phys. Chem. Solids* **27**, 705, 1966.

[23] Sonder, E. and Sibley, W. A., *Phys. Rev.* **129**, 1578, 1963.

[24] Thommen, K., *Phys. Letters* **2**, 189, 1962.

[25] Agullo-Lopez, F., *Phys. Stat. Sol.* **22**, 483, 1967.

[26] Pooley, D., *Proc. Phys. Soc.* **89**, 723, 1966.

[27] Markham, J. J., *S.S.P. Supp.* **8**, 1966.

[28] Fowler, W. B., Editor, *Physics of Color Centers* (Academic Press) 1968. Fowler, W. B., Chapter 2.

[29] Dawber, P. G. and Parker, I. M., *J. Phys. C*, **3**, 2186, 1970.

[30] Mollwo, E., *Z. Physik* **85**, 56, 1933.

[31] Ivey, H. F., *Phys. Rev.* **72**, 341, 1947.

[32] Bartram, R. H., Stoneham, A. M. and Gash, P., *Phys. Rev.* **176**, 1014, 1968.

[33] Townsend, P. D., Clark, C. D. and Levy, P. W., *Phys. Rev.* **155**, 908, 1967.

[34] Seidel, H., Schwoerer, M. and Schmid, D., *Z. Physik* **182**, 398, 1965.

[35] Hopfield, J. J. and Worlock, J. M., *Phys. Rev.* **137**, 1455, 1965.

[36] Fowler, W. B., Editor, *Physics of Color Centers* (Academic Press) 1968. Mahr, H., Chapter 4.

[37] Ueta, M., *J. Phys. Soc. Japan* **23**, 1265, 1967.

[38] Ueta, M., Kondo, Y., Hirai, M. and Yoshinary, T., *J. Phys. Soc. Japan* **26**, 1000, 1969.

4

THERMOLUMINESCENCE

4.1 Basic thermoluminescence theory

Insulators which contain electrons trapped at defect sites are in a metastable condition. The trapped charge can be liberated by heat and will cross the potential barrier and move to a lower energy state with the emission of light. This is called thermoluminescence. The rate at which electrons, or holes, escape from the trap is governed by the vibrational frequency of the charge within the trap Y and the height E of the potential barrier, so that the overall rate of escape is proportional to $Y \exp(-E/kT)$. Very naively we can see that on heating a crystal with trapped electrons the luminescence signal starts at zero (no electrons escaping), increases with increasing temperature (the Boltzmann factor increases) and finally drops to zero at high temperatures when all the initial levels are empty. This maximum in the luminescence is termed a glow peak. Detailed analysis of the curve shape can give information on the trap depth, frequency factor Y, number and type of electron and hole traps, electron mobility and capture cross sections of the various levels. Since we are considering a very sensitive technique which is capable of detecting as few as 10^7 defects in a sample, we are faced with a very complex problem in a real system where there will be a total concentration of 10^{16} or more defects. The theoretical problem is still tractable, since this only requires that we solve the problem with more variables, but detailed comparison with experiment can no longer yield unique solutions for the variables. Historically the theoretical approach has been either to describe a very simple idealized process, or to use a digital computer to predict the possible consequences of changing the variables. Many approximations have also been suggested which correctly estimate the trap depth but not the other parameters. We shall start with the simple model of Randall and Wilkins[1] in which the electrons are thermally excited from a single defect level to the conduction band and radiatively decay to a luminescent site, as is described by Fig. 4.1. We assume that no direct transitions take place from the defect to the recombination centre and

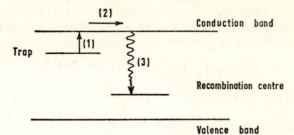

Fig. 4.1 The processes involved in thermoluminescence: (1) thermal excitation; (2) electronic migration and (3) luminescence.

the number of defect sites is small compared with the number of luminescent centres and also the recombination lifetime in the conduction band is small. The rate of depopulation of the defect is

$$\frac{dn}{dt} = -n \, (N_c \, Sv) \, \exp\left(-\frac{E}{kT}\right) + An_c \, (N - n). \qquad (4.1)$$

In the first term n is the electron concentration in the defect, N_c is the density of states in the conduction band, S is the electron capture cross section in the conduction band and v is the electron velocity. The second term allows for the possibility of a back reaction with A as the defect's electron capture cross section. N is the total concentration of the shallow levels and n_c the electron concentration in the conduction band.

As a first simplification we assume that retrapping is negligible. This is a valid assumption if the concentration of traps is small compared with the concentration of recombination centres. Further, if the recombination lifetime τ is very short the electron concentration in the conduction band n_c will remain essentially constant. Loss takes place to the recombination centres or the trapping levels, so in general

$$\frac{dn_c}{dt} = -\frac{n_c}{\tau} + \frac{dn}{dt}. \qquad (4.2)$$

When τ is short, dn_c/dt is negligible compared with the other terms. We may also relate the time and temperature by $T = \beta t$, if we raise the temperature at a constant rate so equation (4.2) becomes

$$\frac{n_c}{\tau} = \beta\frac{dn}{dT}. \qquad (4.3)$$

It is the movement of electrons from the conduction band to the deeper levels which provides the luminescence so n_c/τ is proportional to the light intensity I. In the case of no retrapping ($A = 0$) integration of equation (4.1) provides

$$n = n_0 \exp\left[-\int_{T_0}^{T} \frac{N_cSv \exp[-(E/kT)] \, \mathrm{d}T}{\beta}\right]. \qquad (4.4)$$

This gives the light intensity

$$I = \beta \frac{\mathrm{d}n}{\mathrm{d}T}$$

$$= n_0 N_c Sv \exp\left[-\left(\frac{E}{kT}\right) - \int_{T_0}^{T} \frac{N_cSv \exp[-(E/kT)] \, \mathrm{d}T}{\beta}\right] (4.5)$$

which is a glow curve of the form shown in Fig. 4.2 and is termed

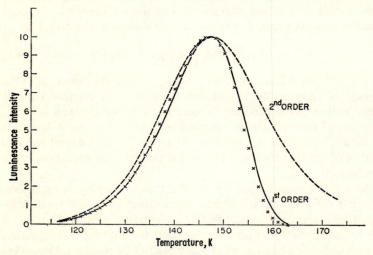

Fig. 4.2 A comparison of experimental data, indicated by crosses ×, for the 147 K glow peak in LiF with simple first and second order kinetics. The heating rate was 1·26°C per minute and the computed curves are for $E = 0.26$ eV and $S = 1.53 \times 10^6$ per second.

'first order' annealing.

If the shallow energy levels also compete for the conduction electrons then we can derive a 'second order' expression for the glow curve which is of the form

$$I = \frac{n_0 N_c Sv \exp[-(E/kT)]}{\left[1 + \int_{T_0}^{T} \frac{N_cSv \exp[-(E/kT)] \, \mathrm{d}T}{\beta}\right]^2} \qquad (4.6)$$

if we assume an equal probability of recombination or retrapping. Fig. 4.2 shows such a glow curve. The main difference from the simple curve is that a greater fraction of the light is produced at temperatures above the peak temperature because the retrapping effectively delays the release of the electrons.

For simplicity most analyses assume the product (N_cSv) can be taken as a single parameter which is temperature independent. Keating[2] pointed out that one might expect $N_c \propto T^{3/2}$, $v \propto T^{1/2}$ and $S \propto \text{fn}(T)$. Then it is only for $S \propto T^{-2}$ that the above estimate is valid. His consideration of alternative exponents for the power law $S \propto T^x$ alters the shape of the glow curve, and from a study of the asymmetry of the curve one may evaluate x.

4.2 Numerical analysis of glow curves

The equations governing the luminescence process do not readily lead to analytically tractable solutions even for simple cases. It is therefore more convenient to ignore the simplifications and compute the resultant glow curve with a digital computer. Such a treatment was performed by Kemmey et al.[3] and we shall describe both his approach and the results for a range of parameters. He assumes a set of defect levels and assigns to each a frequency factor, trap depth, concentration, retrapping cross section and a state of excitation. The last specifies whether the defect is empty, electron- or hole-filled and whether it is in a ground or excited state. One may also allow any of these parameters to be temperature dependent.

At the starting temperature the rate equations are used to calculate the rearrangement of the charges which takes place during a short time interval. The new population conditions are then used for the second calculation at the next temperature. A suitable choice of the time increment and the reiteration rate provides a smooth theoretical glow curve.

Most approximate methods of analysis give a value for the trap depth and the frequency factor. A comparison of the most common methods is given in the paper by Nicholas and Woods[4]. The simplest of these merely assumes that $E \approx 25kT_{max}$, where T_{max} is the temperature of the peak. Since there is no unique pair of values for E and Y that will satisfy this condition, we should examine different pairs of values to see their effect on the curve shape. This is shown in Fig. 4.3. One sees that a deep trap can appear at a low temperature, if the pre-exponential factor is sufficiently large. Comparison with experiment suggests that the curve should have a half-width of about

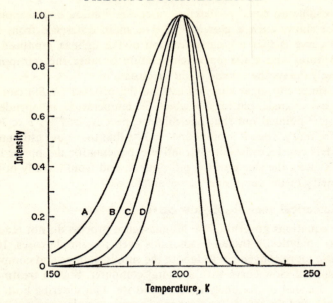

Fig. 4.3 Theoretical glow peaks computed for various pairs of activation energy E and pre-exponential factor Y so that they have a maximum intensity at 200 K. The values used were

	E	Y
A	0·2eV	$6·37 \times 10^2$ second^{-1}
B	0·3	$3·17 \times 10^5$
C	0·4	$1·40 \times 10^8$
D	0·6	$2·30 \times 10^{13}$

40 K if the peak is at 200 K. This means a pre-exponential factor of 10^{12} per second. In this particular case one can see the justification for the simple approximation. It is also tempting to equate this frequency with the lattice vibrational frequency and assume that this is the parameter that governs the electron's escape. However, as was indicated earlier this factor Y is more complex and the simple approximation can be fortuitous. Detailed analysis of the curves provides examples with values of Y even as low as 10^5 per second.

A computational method becomes more valuable when one attempts to determine the effect of different concentrations of shallow traps and recombination centres. For example, if there are insufficient recombination centres, the electron's lifetime in the conduction band must increase as the shallow levels are emptied. If retrapping also occurs, we must compute the product of the trapping

cross section and the concentration of empty energy levels throughout the observation of the glow curve. Computed glow curves are
shown in Fig. 4.4 for the simple case of one defect and one recombination level with equal electron capture cross sections.

When the recombination centres dominate (e.g. their initial concentration is 100 times that of the defect level), the curve has the
asymmetric appearance of the simple case without retrapping. On

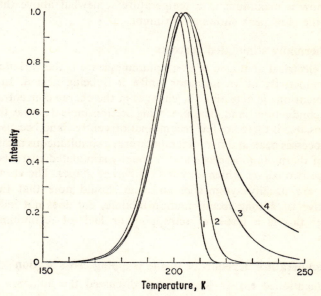

Fig. 4.4 Calculated glow curves for cases where retrapping occurs to
show the curve dependence on the number of recombination
centres. An initial number of traps was chosen as 10^{16} and the
number of recombination centres as (1) 10^{18}, (2) 5×10^{16},
(3) 2×10^{16}, (4) 10^{16}. E was 0·5 eV, Y was 5·8 $\times$ 10^{10} second^{-1}.

the other hand when the two types of centre have similiar concentrations, one finds a very broad glow peak which has a long high-
temperature tail. There is also a small shift in the position of the peak
maximum. For the purposes of estimating the trap depth one can
use the shape of the low-temperature edge of the curve, since this
is insensitive to the amount of retrapping and approximates to exp
$(-E/kT)$ in either case. In practice both these estimates are useful
as one cannot always produce the peaks in isolation, and overlap
causes uncertainties in analysing the curve shape.

The example of equal numbers of shallow and deep traps is particularly important, since this is just the condition that one may achieve by an irradiation or chemical doping where there are equal numbers of vacancies and interstitials, or impurities and compensating vacancies.

Although it is not apparent in the glow curve, the concentration of conduction band electrons increases under these conditions and may also show a maximum at a temperature somewhat above that at which the glow peak shows a maximum.

4.3 Thermally stimulated currents

The electrical analogue of thermoluminescence is the variation of the conductivity of an insulator whilst it is being heated. In this experiment one is detecting the changes in the charge concentration in the conduction band. The preceding section indicates that this is only possible if there are few recombination centres, and because the two processes peak at different temperatures a simultaneous measurement of thermoluminescence and thermally stimulated currents will produce two curves which may have different shapes. The electrical curves are usually symmetric, and one should note that this is indicative of a long recombination lifetime, but does not indicate whether this is a result of retrapping or lack of recombination centres.

4.4 Temperature dependence of the trapping cross section

As mentioned earlier Keating has discussed the influence of a temperature dependent trapping cross section on the shape of the glow curve. Changes in the other parameters can produce similar curve shapes to those in Fig. 4.4, and it is rather difficult to separate experimentally the effects of a limited number of recombination centres and the temperature dependence of the parameters N_c, v or S. If S is simply a function of temperature, one may observe the glow peak in different temperature ranges and hence determine the temperature dependence of S. One can displace the glow peak by varying the heating rate and the peak temperature is related to the rate of rise of temperature by

$$\frac{SN_c v \, kT_{max}}{\beta E} = \exp\left(\frac{E}{kT_{max}}\right) \tag{4.7}$$

In practice this allows one to move the peak position by 20 or 30 K and over this limited range look for temperature dependent parameters. Equation (4.7) has also been used as a means of estimating the trap depth, and an example of this type of analysis is shown in Fig. 4.5 for a glow peak in lithium fluoride.

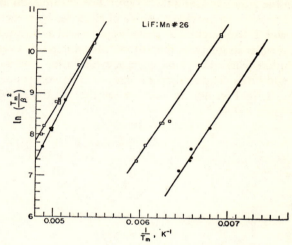

Fig. 4.5 Examples of the plot $\ln\left(\dfrac{T_m{}^2}{\beta}\right)$ versus $\dfrac{1}{T_m}$. The glow curves were obtained either after ultra-violet irradiation, indicated by □, or after X-ray irradiation, indicated by ●.

4.5 Electron and hole untrapping

From the measurement of the luminescence or the electrical current we wish to determine the sign of the released charge. Electrically the currents are rarely sufficient to determine the sign of the Hall co-efficient. Optically there are several choices, and the first of these is to measure the emission spectra for each glow peak. From Fig. 4.5 one can see that if the charge moves from the trap to the luminescent centre via a band, then all electron traps will have one spectrum and this will differ from that produced by hole release. A totally different spectrum may also imply that there is direct charge transfer between a pair of localized levels. When a band is involved it is only necessary to determine the sign of the charge for one defect and all the others are automatically identified. Trivial solutions exist if we can directly associate a glow peak with the annealing, or formation of an electron spin resonance signal or a known optical absorption band. Unless one can do this for several defects there is still the possibility of a mistake if one moves charge from a complex defect or one which contains both a bound electron and a hole, since in this instance release of an electron will allow a hole type defect to appear. An alternative method is to provide a known charge from another defect and observe the changes in the glow curve and the spectra. For example, in an alkali halide it is possible to provide a source of electrons by bleaching the F centres with F band light. Glow curves produced as a result of this are much simpler to interpret than those following X-ray or gamma-ray irradiation where the more energetic ionization liberates both electrons and holes. The method of analysis is demonstrated by the results shown in Fig. 4.6. In this we see that all the glow curves

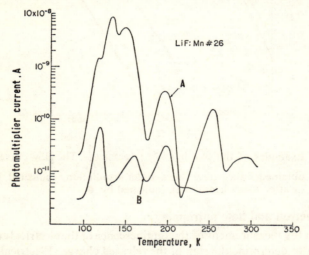

Fig. 4.6 A glow curve *A* produced in LiF : Mn by x-ray irradiation at 80 K and warming to 300 K and the glow curve B produced by recooling the crystal to 80 K and irradiating it with light in the F absorption band. Both measurements were made at 6 K per minute heating rate.

appeared after X-ray irradiation, whereas charge transfer of electrons from the F centres only stimulated the glow peaks at 115, 154 and 194 K. These three defects are thus electron traps.

Non-appearance of the other glow peaks could occur because (i) they are hole traps or (ii) they represent electron trapping defects which are formed during X-ray irradiation but completely anneal out after the loss of the charge. In the example quoted it was also possible to X-ray irradiate *and* bleach the F centre before heating the crystal. The additional electrons increased the fraction of electron traps which were populated, even if they had been formed by the X-ray irradiation, and secondly electron-hole recombination took place at hole traps with a reduction in the intensity of the corresponding glow peaks. The result of this experiment indicated that the peaks at 133 and 147 K were hole traps. Further confirmation was possible as the light intensity was sufficient to measure the emission spectra of the individual peaks. This showed two basic emission spectra.

4.6 The emission spectra

As mentioned before, charge release which proceeds via a lattice energy band will produce the same luminescence spectrum for all defects which release the same charge. Therefore one may classify both the shallow trap and the luminescent centre into either the elec-

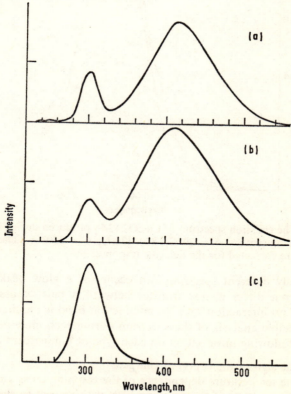

Fig. 4.7 Luminescence spectrum of KCl : Tl under various conditions. At (a) 77 K; (b) 200 K; (c) 273 K. The resolution was reduced in case (c).

tron or hole category. In the alkali halides spectra have been obtained from the more intense glow peaks[5], and examples of the types of spectra are shown in Fig. 4.7. For intense glow peaks, such as that found in magnesium doped calcite, the temperature dependence of the emission peak can also be measured. As is shown in Fig. 4.8 changes in the peak width and fine structure are apparent.

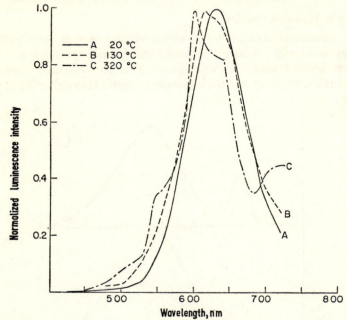

Fig. 4.8 The emission spectrum of $CaCO_3$: Mn produced during thermo-luminescence A = 20°C, B = 130°C, C = 320°C. (The curves are corrected for the detector response.)

A totally different spectrum will occur for a glow peak which results from direct charge transfer between a pair of associated centres if no interaction with the conduction band is required.

In principle analysis of the spectrum during each glow peak can give the following information on each type of luminescent centre:

(i) The sign of the captured charge.
(ii) The temperature dependence of the trapping cross section.
(iii) The position of the trapping level with respect to the Fermi level.
(iv) The class of luminescent centre.

To separate these factors one must be able to run a glow curve from a reproducible starting condition to ensure that the initial number of each type of defect is the same and hence stabilize the position of the Fermi level. Temperature dependence of the capture cross section can be treated, as with S, by moving the temperature of the peak by altering the heating rate.

It should be noted that there are two types of luminescent centre. The first are defects which accept a charge and then are complete, so they cease to act as luminescent centres. The second are recombination centres at which electrons and holes annihilate. The latter may be small in number but can act as luminescence centres so long as both electrons and holes are free. Such centres are also important in photoconductivity measurements.

The full complexity of the problem is both revealed and simplified by the spectral analysis of the light emitted at each stage of the glow peak. In a very elegant experiment by Mattern *et al.*[6] this has been done and the results are fed to a computer which presents them as a three-dimensional view of the emission intensity, wavelength and temperature. This is shown in Fig. 4.9 for a sample of KCl : Tl

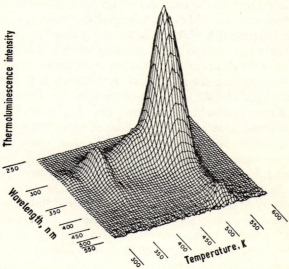

Fig. 4.9 A 'three-dimensional plot' of the thermoluminescence from a KCl : Tl crystal containing approximately 100 ppm Tl$^+$, after it had been exposed to a dose of 3×10^7 rad of Co^{60} gamma rays. The sample was heated linearly at approximately 10°C per minute (Mattern, Lengweiler, Levy and Esser[6]).

irradiated at room temperature with Co^{60} gamma rays for a dose of 3×10^7 rads. The heating rate was $10°C$ per minute.

4.7 The sensitivity of thermoluminescence

Thermoluminescence is particularly sensitive, because one can thermally separate the various defect levels and optically detect them with high efficiency. Although fast heating rates produce a rapid, and therefore more intense, burst of light from the crystal, in order to make an accurate analysis of the curve we also require careful temperature control. This places an upper limit on the heating rate of about $30°C$ per minute. Figs 4.2 and 4.6 show that for a peak at 200 K one might expect a signal over a range of 50 K. The signal level will be adequate for analysis if it emerges from the photomultiplier background noise to a maximum level of, say, 100 times the signal to noise level. With a photomultiplier tube operating at a gain of 10^5 a dark current of 10^{-12} amperes is possible. Since one knows that the quantum efficiency of the photocathode (i.e. photoelectrons emitted per photon $\times 100$) is around 20% and one can collect 20% of the emitted light in a simple arrangement of the apparatus, the total glow curve represents the emission of only 10^7 photons from the sample. More sophisticated experimentation can clearly improve this sensitivity. Relating this peak size to the actual numbers of electrons released from traps is not so simple, because for this we need to know the fraction of the electrons which are lost from the conduction band by radiative processes, the fraction of the light which is reabsorbed within the crystal and the spectrum of the light and the wavelength response of the photomultiplier. Despite these doubts we can estimate that thermoluminescence is capable of detecting as few as 10^9 defect levels in a sample and also separate the various levels. This is a sensitivity unrivalled by any other simple technique. It is for this reason that it has found many applications ranging from personnel radiation dosimeters to systems which are used to study the age and thermal history of archaeological specimens. The obvious application of using thermoluminescence for chemical analysis does not yet seem to be used, possibly because chemical 'purity' is still six orders of magnitude above the lower limit of the system and all samples look very impure.

Chapter 4 References
[1] Randall, J. T. and Wilkins, M. H. F., *Proc. Roy. Soc.* **A184**, 366, 1945.
[2] Keating, P. N., *Proc. Phys. Soc.* **78**, 1408, 1961.

[3] Kemmey, P. J., Townsend, P. D. and Levy, P. W., *Phys. Rev.* **155,** 917, 1967.

[4] Nicholas, K. H. and Woods, J., *Brit. J. Appl. Phys.* **15,** 783, 1964.

[5] Ghosh, A. K., *Appl. Optics* **2,** 243, 1963.

[6] Mattern, P. L., Lengweiler, K., Levy, P. W. and Esser, P. D., *Phys. Rev. Letters* **24,** 1287, 1970.

5

PHOTOCONDUCTIVITY

5.1 Photoconductivity

For an insulator or semiconductor with no impurity or other energy levels in the band gap the number of electrons in the conduction band will equal the number of holes in the valence band and this will remain constant at a given temperature. Holes and electrons recombine at the same rate at which electrons are thermally promoted from the valence to the conduction band. These charge carriers are responsible for intrinsic conductivity.

Illumination with photons of energy greater than the band gap, that is wavelengths shorter than the absorption edge, will excite

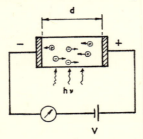

Fig. 5.1 A photconductor in which both holes and electrons are mobile, fitted with end contacts a distance d apart and with an applied voltage V.

electrons across the band gap and thus create further electron-hole pairs which then contribute to the conductivity. For semiconductors the band gap ranges from 0·2 to 2·5 eV for which the threshold wavelengths for intrinsic photoconductivity are 6·2 and 0·5 microns.

This enhancement of the number of charge carriers increases the conductivity of the substance and the phenomenon is called photoconductivity. The basic photoconductive circuit is shown in Fig. 5.1 where light falling on a photoconductor produces electron-hole pairs uniformly throughout the bulk of the photoconductor. Any electrons leaving the photoconductor at the positive electrode are replaced by electrons entering from the negative electrode. Electron-hole pairs combine directly by the electrons dropping back across the band gap.

The photoconductivity is maintained during illumination and ceases almost immediately after the light is shut off.

In real photoconductors the situation is much more complex. Holes and electrons do not contribute equally, since the electrons normally have a much greater mobility and so dominate the photoconductivity. Photoconductivity may be produced by wavelengths λ_i longer than that of the absorption edge λ_g as shown in Fig. 5.2. We should note that the photoconductivity declines again as the photon energy is increased beyond the absorption edge. Photoconductivity

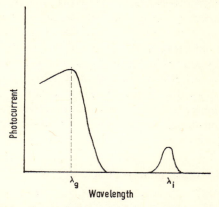

Fig. 5.2 The variation of photocurrent with wavelength for a typical photoconductor. At wavelengths shorter than the band edge λ_g the light is strongly absorbed in a surface layer which enhances recombination of holes and electrons and thus reduces photoconductivity. Impurity levels in the band gap give rise to photoconductivity at longer wavelengths such as λ_i.

produced by longer wavelength radiation λ_i is produced by electrons promoted from impurity levels which always exist in the band gap of real materials, and their distribution and occupancy are of crucial importance in practical photoconductors. The decrease in photocurrent with more energetic photons is a surface effect, because the solid becomes more highly absorbing for $\lambda < \lambda_g$ and the radiation produces a high density of carriers in a very thin surface region of the crystal. For reasons which we will consider below, the surface enhances recombination of electrons and holes, thus removing charge carriers more rapidly than the bulk of the crystal. The efficiency of the photoconductor falls as shown in Fig. 5.2 as the irradiated surface region becomes shallower and the ratio of surface to volume increases. We must also consider additional complications such as space charge and electrode effects.

With modern refinements in the measurement of small currents, photocurrents have been detected in almost every class of substance, including metals and superconductors.

The early history of the field was greatly concerned with seeking new photoconductors, whereas now it is recognized that most substances can display photoconductivity given the correct illumination and measuring methods. Most of the commercially useful photoconductive devices involve semiconductors, but photoconductivity has been measured in substances whose resistivities range from below 1 ohm cm to 10^{18} ohm cm. Another characteristic of importance in photoconductors is the lifetime of the photo-induced charge carriers. Impurity centres in the band gap can greatly modify the band to band recombination process, and the lifetime of the free charge carriers may be greatly increased. The lifetime of photo-induced charge carriers varies from 10^{-13} seconds up to several seconds and, in a particular substance, can be modified by the controlled addition of impurities.

5.2 The Fermi level in photoconductors

In order to specify the occupancy of the electron, or hole, trapping levels we must know the position of the Fermi level. The Fermi-Dirac distribution function gives the probability $f(E)$ of any state of energy E being occupied as

$$f(E) = \frac{1}{1 + \exp\left(\dfrac{E - E_F}{kT}\right)} \tag{5.1}$$

where E_F is the Fermi level, the translational energy of an electron at the top of the distribution at 0 K. The situation in an insulator where E_F lies in the band gap and in which there are a number of localized states is shown in Fig. 5.3. Only localized states below E_F will be occupied at 0 K and at higher temperatures T only those of the order of kT above E_F have a significant probability of occupation.

At energies well above E_F in the tail of the distribution $E - E_F \gg kT$ and equation (5.1) reduces to

$$f(E) = \exp\left(-\frac{E - E_F}{kT}\right) \tag{5.2}$$

which is a Boltzmann distribution.

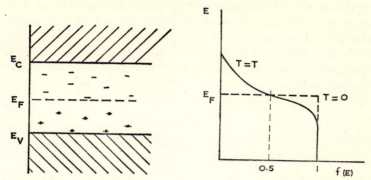

Fig. 5.3 The Fermi level E_F in an insulator which has localized states in the band gap. At 0 K only states below E_F will be occupied and at T K only states less than kT above E_F are likely to be occupied.

The distribution of holes in the valence band of an illuminated photoconductor obeys the same statistics and leads to a hole Fermi level which is not at the same energy as the electron Fermi level. We thus have two quasi-Fermi levels during illumination of a photo-conductor. The electron quasi-Fermi level E_{Fn} is above the normal Fermi level and the hole quasi-Fermi level E_{Fp} is below (Fig. 5.4).

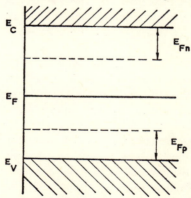

Fig. 5.4 The quasi-Fermi levels for electrons E_{Fn} and holes E_{Fp} in an illuminated photoconductor.

5.3 Gain, mobility and lifetime

The sensitivity of a particular device, with the circuit shown in Fig. 5.1, depends on the number of charges collected by the electrodes per photon absorbed. This ratio G is called the quantum efficiency or quantum gain or photoconductivity gain. The maximum value of this gain is unity and will obviously be less if either carrier is trapped without reaching an electrode.

If the lifetimes of the carriers are τ_n and τ_p for holes and electrons and the transit times are t_n and t_p for the carriers to move between the electrodes, then

$$G = \frac{\tau_n}{t_n} + \frac{\tau_p}{t_p}. \tag{5.3}$$

The mobilities of the carriers μ_n and μ_p are their velocities per unit electric field, so that the transit time for electrons will be

$$t_n = \frac{d^2}{\mu_n V} \tag{5.4}$$

where a voltage V is applied to electrodes of separation d.

Including the hole contribution, equations (5.4) and (5.3) give

$$G = \frac{V(\tau_n \mu_n + \tau_p \mu_p)}{d^2}. \tag{5.5}$$

The increase in conductivity $\Delta\sigma$ produced by illumination is thus

$$\Delta\sigma \propto (\tau_n \mu_n + \tau_p \mu_p). \tag{5.6}$$

In unilluminated insulators the free carriers are few and hence $\Delta\sigma \gg \sigma$ but in semiconductors the reverse applies. The photoconductive effect is hence proportionately easier to detect in insulators.

The mobilities are least in ionic substances because of the high Coulombic interaction between the ions and the charge carriers: μ_n for AgCl is 70 cm^2 volt^{-1} second^{-1} compared with 65000 cm volt^{-1} second^{-1} for InSb. The corresponding values of μ_p are 40 and 1000 cm^2 volt^{-1} second^{-1}. The mobility is also a function of temperature. An extensive table of mobilities and their temperature variations may be found in the book by Bube[1]. As mobility is related to the band structure of the material, it may also depend on crystallographic direction and be influenced by an applied electric field.

In real photoconductors electrons and holes may be captured by imperfections, or traps, with energy levels lying in the band gap. However, if the trap is shallow they may be subsequently released by thermal energy. If only one carrier is likely to be captured the centre is called a trap. If both are easily captured it is a recombination centre and provides an alternative to band-to-band transitions for recombination of electrons and holes. Traps prolong the decay of photoconductivity when the light is switched off, and hence photoconductivity measurements enable us to study the nature of these defects and imperfections in insulators.

Both kinds of centres may reduce the sensitivity of the photoconductors by removing charge carriers. We will now consider the dynamics of the photo-conductive processes when a photoconductor is illuminated.

5.4 Recombination without traps

At the onset of illumination the number of photoconductive free carriers is low and so is the recombination rate. The recombination rate increases till the carrier concentration reaches saturation when creation and recombination are equal. The simplest recombination process depends on the concentration n of only one type of carrier, say electrons, with the hole concentration p remaining relatively constant

$$\dot{n} = - an. \tag{5.7}$$

By analogy with chemical reactions this is called a monomolecular process[2] which at equilibrium gives

$$\dot{n} = L - an_0 = 0 \tag{5.8}$$

where L charge carriers are produced by illumination per unit volume per unit time, and n_0 is the steady state electron concentration

$$n_0 = \frac{L}{a}. \tag{5.9}$$

The photoconductive current is thus linearly proportional to the illumination for monomolecular recombination.

The next simplest recombination process depends on both holes and electrons, assumed to be equal in number,

$$\dot{n} = - bnp \tag{5.10}$$

which at equilibrium gives

$$\dot{n} = L - bn_0p_0 = L - bn_0{}^2$$
$$n_0 = \left(\frac{L}{b}\right)^{\frac{1}{2}}. \tag{5.11}$$

This bimolecular reaction gives a photocurrent which is proportional to the square root of the light intensity. Experimentally one finds the photocurrent varying as a power of light intensity from a half to one or more.

The decay rate of the photoconductivity comes from the integrations of equations (5.7) and (5.10). For the monomolecular case

$$n(t) = \frac{L}{a}e^{-at} = n_0 e^{-\frac{t}{\tau}} \tag{5.12}$$

and the bimolecular case

$$n(t) = \frac{n_0}{1 + tbn_0} = \frac{n_0}{1 + \dfrac{t}{\tau_0}} \tag{5.13}$$

where $n = n_0$ and $\tau = \tau_0$ at $t = 0$.

The carrier concentration thus drops by half in a time

$$t_0 = \frac{1}{bn_0} = \frac{n_0}{L}. \tag{5.14}$$

Thus the response time is inversely proportional to the sensitivity of the photoconductor, and it is for this reason that photoconductive light meters designed for low light intensity respond sluggishly. The

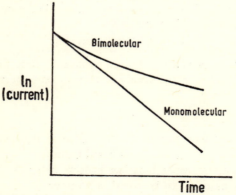

Fig. 5.5 The decay of photocurrent after illumination is removed. For monomolecular recombination the carrier concentration decreases exponentially and log current versus time gives the straight line. The slope of the bimolecular recombination curve decreases with time.

two processes can be distinguished by plotting the log of the current against the time after the light is removed, as shown in Fig. 5.5. A monomolecular recombination gives a straight line, and a bimolecular process gives the other curve. Most experimental results lie between these two curves.

An example of a bimolecular system is an insulator of great purity and perfection which will have few levels in the band gap and few carriers in the dark. Equal numbers of electrons and holes are photo-excited when the insulator is illuminated by photons of greater energy than the band gap $hv > E_c - E_v$. Recombination will be bimolecular (Fig. 5.6), as it depends on the equal concentrations of

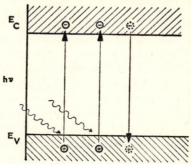

Fig. 5.6 Bimolecular recombination produced by illumination of energy greater than the band gap. Recombination depends on equal numbers of holes and electrons, in the absence of traps.

both holes and electrons.

An insulator with many occupied levels low in the band gap, illuminated by photons with energy $hv < E_c - E_v$, capable of exciting electrons only from these levels to the conduction band (Fig. 5.7),

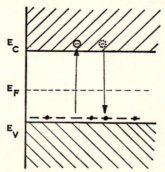

Fig. 5.7 A small number of electrons excited from numerous occupied levels near the valence band by light of lower energy than the band gap. Recombination depends only on the number of electrons, since the change in the number of occupied levels is small at low levels of illumination. The recombination is monomolecular under these conditions.

will exhibit monomolecular recombination at low light levels. The reason is that the low excitation rate causes little change in the concentration of empty centres and the recombination depends only on the number of electrons in the conduction band. As the light intensity increases the empty centre concentration increases and the process becomes less monomolecular and more bimolecular.

5.5 Demarcation levels

The nature and distribution of the defects is our main concern here, but before considering the specific physical nature of imperfections we must look briefly at the phenomenological approach that is used[3], particularly to discuss devices. This approach concentrates on the location of centres in the gap relative to the Fermi levels and uses a demarcation level to divide centres into recombination or trapping centres.

For a centre to act as a recombination centre it must capture a carrier and hold it long enough to capture a carrier of opposite sign. The two excited carriers then combine at the centre, releasing the excess energy as photons, phonons or in an Auger-like three-body event in which a third hole or electron takes up the energy.

Centres are distributed throughout the band gap, particularly in

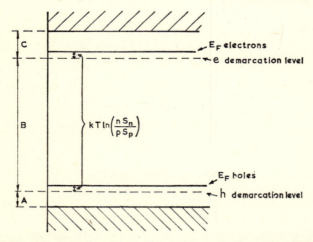

Fig. 5.8 The demarcation levels for holes and electrons. Centres in region A will act as hole traps and electron recombination centres. Centres in region B will act as recombination centres for both holes and electrons. Centres in region C will act as electron traps and hole recombination centres.

impure or polycrystalline materials. Those near the bottom of the conduction band are most likely to trap electrons. Those near the top of the valence band are most likely to trap holes. Somewhere in between, probabilities of capture are comparable and recombination is most likely. These are called recombination centres.

The classification must not be taken too rigidly as recombination can occur at any of the traps; the probability of recombination is, however, much lower than the probability of trapping in the centres classified as trapping centres. A suitable demarcation level may be taken at which an electron has equal chances of free hole capture or thermal excitation into the conduction band. Centres closer to the conduction band than this are traps. Centres closer to the valence band are electron recombination centres. A similar criterion may be taken for hole traps near the valence band.

Traps in regions A and C of Fig. 5.8 are thus in thermal contact with the appropriate band and their occupancy is determined by a Boltzmann factor. Since the cross section S will vary with the type of centre, each type of centre has its own demarcation level. A particular type of centre will have an electron demarcation level at an energy E_1 below the conduction band given by

$$pvS_p = S_n v N_c \exp\left(-\frac{E_1}{kT}\right) \tag{5.15}$$

where N_c is the density of states in the conduction band, S_n is the free electron capture cross section for a hole occupied centre, S_p is the free hole capture cross section for a centre already occupied by an electron and v is the electron velocity.

The centres will act as recombination centres when (region B, Fig. 5.8)

$$pS_p \gg S_n N_c \exp\left(-\frac{E_1}{kT}\right), \tag{5.16}$$

that is, for energies well above

$$E_1 = kT \ln\left(\frac{pS_p}{S_n N_c}\right). \tag{5.17}$$

They will act as electron trapping centres for energies well below E_1. Similar expressions apply for hole demarcation levels.

The electron demarcation level is displaced from the electron Fermi level and the hole demarcation level from the hole Fermi level by an energy[1]

$$kT \ln\left(\frac{nS_n}{pS_p}\right).$$

The occupation of regions A and C is thus determined by temperature and the occupation of region B by the recombination kinetics of the insulator.

5.6 Capture cross section

For recombination of holes and electrons at a recombination centre the probability of recombination depends on the thermal velocity v of the carrier in the appropriate band and the number density N of recombination centres. The free carrier lifetime is given by

$$\tau = \frac{1}{vSN} \tag{5.18}$$

where S is the capture cross section.

Charged centres will interact coulombically with the charge carriers and either attract or repel them. We can therefore estimate S from the distance of approach r at which the coulombic binding is equal to kT.

$$kT = \frac{e^2}{\varepsilon r} \tag{5.19}$$

where ε is the dielectric constant of the insulator. The capture cross section is then

$$S = \pi r^2 = \frac{\pi e^2}{k^2 T^2 \varepsilon^2} \tag{5.20}$$

which for $\varepsilon = 10$ and $T = 300$ K gives

$$S \simeq 10^{-12} \text{ cm}^2.$$

Coulombic repulsion can reduce the cross section to as low as 10^{-22} cm². An uncharged centre is of approximately atomic dimensions and has a cross section of some 10^{-15} cm².

The range of N varies trom 10^{12} cm⁻³ for perfect pure crystals to 10^{19} for highly imperfect crystals. The free lifetime thus varies from 10^3 to 10^{-14} seconds in this calculation. A typical value is $\tau = 10^{-7}$ seconds. The same considerations apply to trapping centres, but once equilibrium has been reached these will fill by capture and be thermally emptied at the same rate. Hence only in non-equilibrium situations do we have to take them into account.

For many materials the capture cross section for holes is greater than that for electrons. This can enhance sensitivity by removing carriers that would otherwise be available for recombination. In most photoconductive devices such specific traps are introduced to increase sensitivity, usually at the expense of response time. For example, pure cadmium sulphide has hole and electron lifetimes of 10^{-6} to 10^{-8} seconds. The stoichiometry of cadmium sulphide may be changed to give additional cadmium vacancies which preferentially capture holes, thus increasing the lifetime of the free electrons to some 10^{-2} seconds. The cadmium sulphide is hence converted into a much more sensitive photoconductor. The process is called sensitization or activation and is most effective at low light levels. At high levels of illumination the concentrations of holes and electrons are no longer small compared with the number of centres; n and p become almost equal again and the defect induced sensitivity is lost.

5.7 Supralinearity

Over a small range of illuminating intensity a sharp increase in photoconductive current is frequently found. The photocurrent may go up as a high power of the illuminating intensity compared with the linear variation found on both sides of this step. This behaviour is called supralinearity or superlinearity. The reason for this behaviour is that the shift of the Fermi level with illumination brings centres into

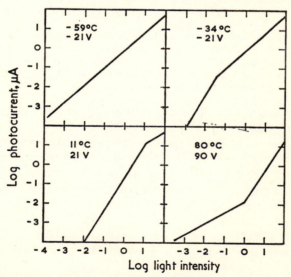

Fig. 5.9 Log photocurrent against log illumination intensity for CdSe at $-59°C$, $-34°C$, $11°C$, and $80°C$, showing the slope changes that occur due to the onset of supralinearity.

play which exert a sensitizing effect on the photoconductor. This behaviour was found by Smith[4] in cadmium selenide and has since been found in numerous other substances[1]. Some results found by Bube[5] on CdSe are shown in Fig. 5.9. At low temperatures the slope of the log photocurrent versus log illumination curve is less than one over the entire range of illumination, as shown on the − 59°C curve. As the temperature increases a slope greater than one occurs at low light levels, as seen on the − 34°C curve. With further increase in temperature (the 11°C curve) the onset of this greater slope region moves to a greater illumination level. At even higher temperatures (see the 80°C curve) a third region with a slope less than

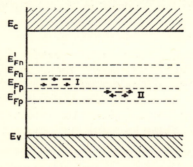

Fig. 5.10 At low light intensities only centres I lie between the hole and electron Fermi levels. As illumination increases E_{Fn} and E_{Fp} move apart to E'_{Fn} and E'_{Fp}, thus encompassing levels II which are predominantly hole traps. This increases the photoconductivity as electrons are left free for a longer period.

unity begins at the low illumination side and progressively moves to higher illumination as the temperature is raised.

Temperature and intensity of illumination produce opposing effects – the higher the temperature the higher is the illumination required to produce the slope change in the photoconductive curve. In Fig. 5.10 at low light levels only the recombination centres I lie between the hole and electron steady state Fermi levels. As illumination increases E_{Fn} and E_{Fp} move further apart to E'_{Fn} and E'_{Fp} and embrace the centres II. The centres I with equal cross sections for electrons and holes of say 10^{-15} cm² and a density of 10^{15} cm⁻³ will allow an electron and hole lifetime of some 10^{-7} seconds[3].

The levels II are predominantly hole traps as they are sufficiently below the steady state hole Fermi level. They are also much more numerous than the class I centres. Thus as illumination increases at a given temperature and E_{Fp} moves down through the II centres, the difference in cross section for electron and hole capture by the II centres preferentially removes holes. The lifetime of the electrons and hence the photoconductivity increases and produces the phenomenon of supralinearity. When all the centres II lie between the hole and electron demarcation levels the slope of the photoconductivity curve, Fig. 5.9, reverts to approximately linear behaviour.

The shift of the supralinear region with temperature is of course due to the fact that the separation of the Fermi levels from the bands is temperature dependent as in equation (5.17). At a given illumination level E_{Fp} moves linearly away from the band edge with absolute temperature. Hence the hotter the crystal the more light is required to encompass the II levels, as is experimentally observed. A fractional change in illumination ΔI is required to balance a fractional change in temperature ΔT in order to keep the onset of supralinearity fixed. Differentiating the expression for the distance of the Fermi level from the band edge gives[3]

$$\frac{\Delta I}{I} = \frac{(E_{Fp} - E_v)}{kT}\frac{\Delta T}{T}. \tag{5.21}$$

One may use this simple model of supralinearity to calculate[1] the position of the Fermi levels relative to the bands, from the temperature of the slope change and the conductivity of the material.

5.8 Space charge limited currents

For a hot cathode emitting electrons into a vacuum an electron space charge may accumulate near the cathode and limit emission[2]. The space charge limited current density is[3]

$$j = 2\cdot3\frac{V^{3/2}}{d^2}\mu \text{ A cm}^{-2} \tag{5.22}$$

where V volts is applied between the plane cathode and a plane anode a distance d away.

If we replace the vacuum by an insulator or semiconductor, the electrons may be injected by a low work function cathode and we can approximate the situation to a capacitor of plate spacing d where

$$Q = CV = \frac{\varepsilon V}{4\pi d} \times 10^{-12} \text{ coulombs cm}^{-2}. \tag{5.23}$$

The transit time between cathode and anode is

$$t_t = \frac{d^2}{V\mu}$$

and hence the current is

$$j = \frac{Q}{t_t} = \frac{\varepsilon V^2 \mu}{4\pi d^3} \times 10^{-12} \text{ A cm}^{-2}$$

$$j = \frac{\varepsilon \mu V^2}{d^3} 10^{-13} \text{ A cm}^{-2}. \tag{5.24}$$

which is much lower than the space charge limited current of a hot cathode emitting into a vacuum. The reduction of conductivity of the solid compared with a vacuum is due to the reduced electron velocity in the solid. In the presence of traps the current will be further reduced by the ratio of free to trapped electrons. Such currents have been observed, and a review of the mechanisms has been given by Wright[6], who has proposed the term dielectric diode for an insulator having an electron-injecting cathode contact and an electron-collecting anode contact. The problems of injecting and collecting currents in non-metallic solids are much greater than those encountered in vacuum devices. We will discuss these contact difficulties later. Electrons may be injected directly into the conduction band and moved through the crystal by an electric field. At low fields carrier

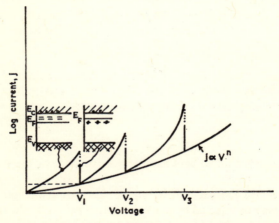

Fig. 5.11 As the voltage is steadily increased to V_1 across the CdS, electrons are injected and the current increases. The voltage is now held steady at V_1 and the current rapidly drops to the lower $j \propto V^n$ curve as most of the electrons are trapped. The current j_1 is the space charge limited current at V_1. Repeating the process to V_2, V_3 ... produces a similar series of peaks. The Fermi level rises at each stage, corresponding to an increase in the number of carriers.

diffusion predominates and gives an exponential dependence of current on voltage. At higher fields drift takes over, giving a square law dependence.

As the voltage is increased across the contacts the current increases at a rate much greater than that predicted by equation (5.24). Typical[1] log current versus voltage curves are shown in Fig. 5.11. The surge of current due to electron injection as the voltage increases is quickly reduced as trapping occurs when the voltage is held steady. The $j \propto V^n$ curve to which the space charge limited current returns at each steady voltage stage is steeper than predicted by equation (5.24) and the current is less. The raising of the Fermi level as the number of free carriers is increased accounts for the more rapid increase of j with V and the trapping accounts for the fewer carriers available to carry the equilibrium current. Trap-free conditions can be approximated in fast pulse measurements, of short duration compared with the trapping time, and give results close to those predicted by equation

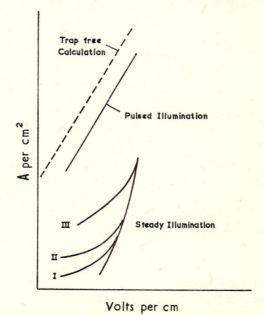

Fig. 5.12 Pulse measurements of photoconductivity on CdS produce results which approximate those expected in the absence of traps. The lower photoconductivity under steady illumination is caused by trapping which does not have time to occur in the fast pulse measurements.

(5.24). Fig. 5.12 shows results of space charge limited and pulsed measurements[7] on CdS both in the dark and at three levels of illumination. The pulsed measurements are much closer to theoretical trap-free conditions than are the other measurements.

The most significant region of the current voltage relationship for an insulator or semiconductor is confined to the triangular region of the log V : log j curve shown in Fig. 5.13. The triangle is bounded by

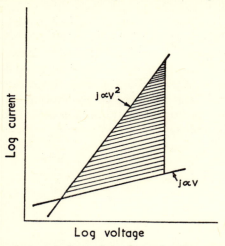

Fig. 5.13 Log current versus log voltage for an insulator or semiconductor. Results are confined to the triangle bounded by $j \propto V$, $j \propto V^2$ and the vertical line corresponding to the limit of trap filling

the ohmic curve ($j \propto V$), the trap-free space charge limited curve ($j \propto V^2$) and the space charge limited curve with filled traps. From an analysis of this last curve the distribution of the traps in the band can be found[8], and from the intersection of the last two curves the trap density N_t may be found from

$$N_t = \frac{CV_t}{edA} \qquad (5.25)$$

where V_t is the intersection voltage, d is the interelectrode spacing and A is the cross sectional area.

5.9 Electrode effects

The need to make effective electrical contact with photoconductors is a difficulty not encountered with many other methods of studying insulators. Concentration gradients of carriers and space charge will always exist near the contacts. The nature of the contact material and its Fermi level will have an influence as will the state of the surface. For example, electron bombardment to remove surface films may enable good contact to be made to a surface that otherwise presents problems due to a surface barrier. The photoconductivity will be influenced by the contacts and in some cases contact effects may dominate the measurements. We will now consider the basic types of contacts[9] that can be made between metal electrodes and semi-conductors or insulators.

Neutral contact

Both the charge transferred into a semiconductor or insulator and the distribution of charge within it are predominantly controlled by the work function of the metallic contacts. If this work function is the same as that of the insulator or semiconductor to which contact is made, there is no electron or hole flow across the interface and no change occurs in either metal or non-metal (Fig. 5.14). Such a

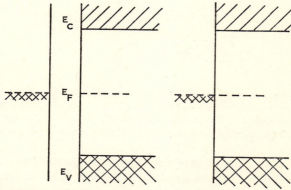

Fig. 5.14 Contact between a metal and a semiconductor or insulator both of which have the same Fermi energy E_F and which are free of any surface contamination. No change occurs on making contact.

contact is called a neutral contact and obeys Ohm's Law up to a saturation current as the non-metal remains homogeneous up to the contact boundary. Such contacts are rare, as it is difficult to find two matching materials, and the problems of contacting them without surface contamination can be considerable.

Blocking contact

If the metallic work function exceeds that of the non-metal, electrons flow from the higher to the lower Fermi level (Fig. 5.15) to equalize the Fermi levels at the point of contact. Here the electrons flow from the non-metal to the metal, leaving an electron-depleted region of depth D in the non-metal which now has positive space charge. The bands bend up as they approach the interface because at equilibrium the bands away from the contact are at their normal bulk levels.

At a distance D from the interface (Fig. 5.15) the non-metal has

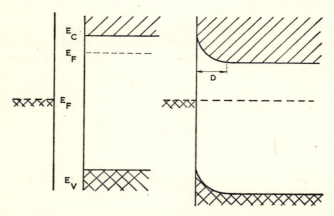

Fig. 5.15 Contact between a metal and an insulator or semiconductor where the work function of the metal is greater than that of the non-metal. Electrons flow into the metal leaving an electron depleted region, of depth D, with a positive space change.

reverted to its normal bulk behaviour. The positive space charge region is spread over this depth D in the non-metal whereas in the metal any charge can be taken as concentrated on the surface, leaving the metal unchanged in depth. A barrier now exists at the surface, and the contact is called a blocking contact. The emission current is saturated as in a vacuum photocell. If the concentration of electrons in the surface region is very low the high resistance of this region may dominate the conductivity regardless of photo-induced changes in the bulk of the material. Applying a positive potential V to the metal narrows the boundary layer and reduces its resistance. A negative voltage V on the metal enlarges the boundary layer and increases its resistance. The contact thus acts as a rectifying contact.

In both cases the current is given by

$$j = C \exp \left(\frac{eV}{kT} - 1 \right) \qquad (5.26)$$

where C is a constant which varies for forward and reverse biasing and the barrier is never thin enough for tunnelling to occur. Examples of solid state devices which operate on blocking contacts are selenium or copper oxide rectifiers and back biased p–n junctions.

Injecting contact

We have so far ignored the effect of tunnelling through thin barriers on the electrical characteristics of our contacts. We have also assumed an abrupt change at the interface. It can be advantageous to modify this abrupt surface step by diffusing impurity atoms into the insulator or semiconductor. For example, diffusing donor atoms into an n-type photoconductor will produce many free electrons at the interface which greatly thin the blocking barrier, and electrons easily tunnel through the barrier. Thus a blocking contact can be converted into an injecting contact, and the restrictions on the surface barrier height and hence the work function can be reduced. We then have a wider choice of suitable contact metals.

Contacts and illumination

As charge carriers are excited by light, the quasi-Fermi levels for the holes and electrons move relative to the Fermi level of the metal contacts. Under intense illumination the movement may be so great that a quasi-Fermi level of the photoconductor passes from one side of the metal Fermi level to the other and so changes the nature of the contact. The situation is complicated if more than one type of centre participates, but we will here confine ourselves to simpler cases which indicate the main electrical characteristics of photoconductors with electrodes under moderate illumination.

With blocking electrodes and similar mobilities of both charge carriers in the photoconductor shown in Fig. 5.1, the photocurrent saturates when the applied voltage is increased under steady uniform illumination. The initial current increase is due to enhancement of the drift of carriers with increasing voltage. The time available for recombination is hence reduced, more carriers reach the electrodes and the photocurrent is greater. At saturation all charge carriers are collected at the electrodes before they have time to recombine. This gives us a measure of the quantum yield (G electrons per incident photon) of the photoconductor under the given illumination. If the transit times are t_n and t_p and the lifetimes τ_n and τ_p for electrons and holes

$$G = \frac{\tau_n}{t_n} + \frac{\tau_p}{t_p}. \tag{5.27}$$

If the holes have a much lower mobility they will give rise to a positive space charge near the cathode and concentrate the field in this region. The rest of the photoconductor having a much reduced field will have a much greater recombination rate and therefore a much reduced contribution to the photocurrent. Light will be effective only in the cathode region. The difficulty can be overcome by changing from blocking to injecting contacts.

We have so far generally assumed that charge carriers are uniformly excited throughout the bulk of the photoconductor. This in practice is not usually true. Both the illumination and the photoconductor can be inhomogeneous. Light is absorbed according to $I = I_0 e^{-\mu t}$ (see Chapter 3), so most light is absorbed near the surface if the absorption coefficient is large. Surfaces effect recombination of holes and electrons. They do so more effectively when the absorption is greater. The variable nature of surfaces must therefore be considered. A review of inhomogeneous effects has been made by Heijne[10], including the effects of gas adsorption. We will confine ourselves to only two such effects, the photovoltaic effect, based on a deliberately inhomogeneous photoconductor, and the Dember effect, due to high absorption of light and therefore inhomogeneous illumination of the photoconductor.

5.10 The photovoltaic effect

The surface barrier layer produced at an interface between a metal and semiconductor is shown in Fig. 5.15. Under illumination such a junction generates a potential difference because of the bending of the bands near the surface. A pair of charge carriers produced by an incident photon are separated by the field in the barrier layer. The mechanism is the same for a p–n junction in a semiconductor where the work function differs on each side of the junction. The region of effectiveness extends about a diffusion length outside the p–n junction, as carriers which can reach the junction region before they recombine will be effected. As the mobilities of holes and electrons may differ the effective region may be asymmetric with respect to the p–n junction. The greater the intensity of the illumination the more electron-hole pairs will form and be separated and the greater will be the potential difference across the junction. This photovoltaic effect has been used to operate photographic exposure meters and solar batteries. The open circuit voltage developed is easily shown[10] to be

$$V = \frac{kT}{e} \ln \left(1 + \frac{g_p}{g_t} \right) \tag{5.28}$$

where g_p and g_t are the production rate of photo-excited and thermally excited electron-hole pairs.

Photovoltaic effects many times greater than those predicted by equation (5.28) and up to 100 V have been found in a number of substances. Evaporated films of cadmium telluride[11] and single crystals of zinc sulphide[12] have shown this anomalous photovoltaic effect. Such large voltages are many times the band gap and must arise by the internal series connection of many barrier layers.

5.11 The Dember effect

If the mobilities of electrons and holes are different they will diffuse at different rates and give rise to a voltage gradient in the direction of the illumination. This is the Dember effect[13]. If the photoconductor is illuminated through a transparent electrode by light which is strongly absorbed, only a thin surface layer will be excited. If the electrons are more mobile than the holes, they will diffuse inwards more rapidly, thus making the transparent surface electrode positive with respect to an electrode on the back of the crystal. If the back and front electrodes are electrically connected the current is due to both diffusion and field induced currents,

$$j_n = -eD_n \frac{\mathrm{d}n}{\mathrm{d}x} - e\mu_n n F_x$$

$$j_p = -eD_p \frac{\mathrm{d}p}{\mathrm{d}x} + e\mu_p p F_x \tag{5.29}$$

where D_p and D_n are the diffusion constants, F_x the electric field, n and p the hole and electron concentrations and μ their mobilities. The direction of the Dember voltage indicates which carrier is more mobile, but because of the space charge usually present near the surface electrode there are difficulties in making really quantitative measurements using this effect.

5.12 Photoconductive defect studies

In photoconductive measurements we can change the intensity and wavelength of illumination, and we can control the temperature of the photoconductor. The range of responses varies enormously from material to material, and large differences may occur between different samples of the same material. In general, however, an increase of illumination further separates the quasi-Fermi levels, thus increasing the recombination region at the expense of the two-trap regions, Fig. 5.8. If significant new states are thus embraced, the change in characteristics of the photoconductor gives information on their location in the gap. The recombination region may be narrowed by increasing the temperature.

Such methods do not yield unique indentification of the nature and concentration of centres involved and are normally supplemented by spin resonance, thermoluminescence and similar techniques discussed in other chapters. Specifically photoconductive methods which are useful are[1]:

(a) The temperature variation of the free carrier concentration is shown by measurements of the Hall constant or dark conductivity.

(b) Low temperature photoconductivity excitation bands may be found from the variation of photoconductivity with illuminating wavelength.

(c) Carrier lifetime depends on the carrier density under recombination centre conditions. The quasi-Fermi levels depend on the number of carriers excited and determine the level occupation. The ionization energy of a centre may hence be found from lifetime measurements.

(d) Thermally stimulated currents are analogous to glow curves, which have been discussed already. The temperature at which the low-temperature stimulated carriers are released gives the ionization energy of the defect.

(e) The field variation of space charge limited currents gives information on both the location and concentration of levels as discussed in § 5.8.

(f) We may use the photoconductive response to separate absorption processes which fall in the same spectral region.

(g) Phase transformations, stacking faults and the degree of crystallinity of solids have been studied by observing their effects on photoconductivity.

(h) Diffusion lengths and lifetimes can be measured using test probes and fine spots of light, or the decay of photocurrents.

5.13 Conclusion

Photoconductive measurements use well developed and straightforward techniques such as small current measuring methods. Against this advantage must be set the fact that the results do not usually provide unique indentification of the defect that is operating in a particular case, without other supporting evidence. When used in conjunction with other methods, such as thermoluminescence, photoconductivity can, however, provide valuable additional information. Although photoconductivity may not uniquely identify a type of defect, a particular type of defect can greatly affect photoconductivity in specific ways, and it is this which makes the phenomenon so useful in a variety of applied fields and for an ever increasing number of practical devices. The use of p–n junction silicon solar cells to power space vehicles is one of the best publicized photoconductive devices but is not the most important. Radiation detectors from the infra-red to nuclear gamma rays are increasingly used in solid state photoconductive devices. Their low cost and small size compared with photomultiplier tubes and ionization chambers are considerable advantages. Xerographic copying processes, photographic exposure meters, the vidicon camera tube, and even the photographic process itself involve photoconductivity. The speed and efficiency of conversion of optical to electrical energy and wavelength discrimination make photoconductive devices essential to modern technology.

Chapter 5 References

[1] Bube, R. H., *Photoconductivity of Solids* (Wiley) 1960.

[2] Heijne, L., *Philips Tech. Rev.* **27,** 47, 1966.

[3] Rose, A., *Concepts in Photoconductivity and Allied Problems* (Interscience) 1963.

[4] Smith, R. W., *RCA Rev.* **12,** 350, 1951.

[5] Bube, R. H., in *Photoconductivity Conference* edited by R. G. Breckenridge, B. R. Russell and E. E. Hahn (Wiley) 1956, p. 575.

[6] Wright, G. T., *Solid State Electronics* **2,** 165, 1961.

[7] Smith, R. W. and Rose A., *Phys. Rev.* **97,** 1531, 1955.

[8] Rose, A., *Phys. Rev.* **97,** 1538, 1955.

[9] Heijne, L., *Philips Tech. Rev.* **25,** 120, 1964.

[10] Heijne, L., *Philips Tech. Rev.* **29,** 221, 1968.

[11] Pensak, L., *Phys. Rev.* **109,** 601, 1958.

[12] Ellis, S. G., Herman, F., Loebner, E. E., Merz, W. J., Struck, C. W. and White, J. G., *Phys. Rev.* **109,** 1860, 1958.

[13] Dember, H., *Phys. Z.* **32,** 554, 1931 and **33,** 207, 1932.

6

INDIRECT MEASUREMENTS
OF DEFECTS

6.1 Indirect measurements of defects

In the preceding chapters we considered some properties of isolated point defects in a perfect lattice. We must also acknowledge that real crystals contain large concentrations of extended defects in the form of grain boundaries, dislocations and surfaces. Such features are well understood and common to all crystalline materials, and there is no conceptual problem in considering the interactions between point and extended defects. However, because a large region of crystal is involved in these reactions the powerful analysis techniques of ESR or optical absorption may not be applicable, so there is often more doubt about a particular interpretation of the experiments. Extended defects cannot be ignored because they play a major role in the rate of defect production and control such processes as the development of photographic images or the changes in shape and strength of materials used in nuclear reactors.

We have also included in this chapter some brief discussion of less sensitive techniques which are used to study defects. We should realize that with defect studies in insulators and semiconductors we have the advantage of experimental techniques which distinguish between different defects. None the less all the measurements available for defect studies in metals (lattice parameter, volume, mechanical strength, resistivity, and stored energy) are also useful for non-metals. Many of these observed features are summations of effects from all types of defects in the solid, so are of limited value. For example, resistivity measurements can be related to electron scattering at specific defects only by indirect evidence from the analysis of annealing stages, theoretical estimates of formation and diffusion energies and radiation displacement thresholds. Some of these points will be expanded in this chapter and in Chapter 8.

6.2 Dislocation effects

Unlike most of the defects we have so far discussed, a dislocation is a line imperfection. An edge dislocation is shown in Fig. 6.1 which is an atom plane terminating at the plane AB. Such a dislocation could be produced by attempting to shear the crystal at the plane

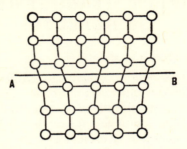

Fig. 6.1 An edge dislocation with an atom plane terminating at the slip plane AB.

AB, which is called a glide or slip plane. Dislocations move at a much lower stress than the yield stress of the perfect lattice and are important in understanding the mechanical properties of real solids[1, 2]. The other principal type of dislocation is the screw dislocation (Fig. 6.2), which we can imagine to be produced by attempting to

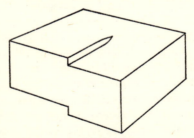

Fig. 6.2 Screw dislocation in a cubic lattice.

twist the crystal lattice. The region near both types of dislocation will be strained and therefore different from the undistorted lattice. This leads to elastic interactions between dislocations: for example, in Fig. 6.3 two edge dislocations of opposite sign on nearby but not overlapping parallel glide planes are able to reduce the elastic strain by combining, and in so doing produce a number of vacancies. If, unlike Fig. 6.3, the half planes initially overlap, then the dislocations, on moving together, give rise to a number of interstitial atoms. This is one example of dislocations changing the number of point defects in a crystal. A further example is the motion of a jog. A jog is illustrated in Fig. 6.4. If the jog, which in its simplest form is a

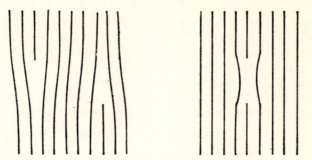

Fig. 6.3 Combination of two edge dislocations of opposite sign on nearly parallel glide planes to produce vacancies.

one-atom step in a dislocation line, moves towards A, vacancies are absorbed at the jog, whereas motion of the jog towards B enables it to absorb interstitials. The role of jogs in the adsorption or emission of point defects is a very important one particularly in adjusting the

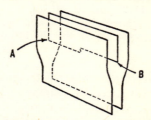

Fig. 6.4 A jog in an edge dislocation. Motion of the jog towards A absorbs vacancies. Motion towards B absorbs interstitials.

equilibrium number of defects as the temperature of the crystal changes. More complicated jog configurations[3] are possible and can occur during slip, but their accumulation into a multi-atom step, or superjog, is unlikely to be stable in metals. The elastic forces between jogs, being mutually repulsive, do not favour jog accumulation.

Impurity atoms or vacancies tend to collect at dislocations, since large impurity atoms, for example, will find it energetically favourable to occupy the comparatively open space at the centre of an edge dislocation. One would not expect this interaction between point defects and screw dislocations, which do not have the volume change associated with edge dislocations. However, interactions other than elastic ones occur, and screw dislocations are also decorated by impurities. The decoration of dislocations in silicon and germanium by copper diffused in from the surface has been used in some elegant experiments by Dash[4]. The semiconductors are transparent to infra-red light, and the copper collected along the dislocation networks is clearly visible when the crystals are viewed in transmission using an infra-red image tube and a microscope.

A dislocation line cannot end within the crystal and must emerge through the surface, unless it forms a closed loop. The region around

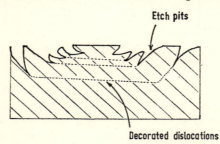

Fig. 6.5 Etch pits in (112) surface of silicon, connected by copper-decorated dislocations. An infra-red image tube and a microscope enable these to be seen in infra-red transparent semiconductors.

a surface-emergent dislocation has a different chemical activity and may be preferentially attacked by a suitably chosen etch, a method which has been extensively used as a means of counting surface-emergent dislocations. One cannot, however, always assume a one-to-one correspondence between etch pits and dislocations without first calibrating the system for the particular etchant which is used. Dash etched surfaces of his copper-doped silicon crystals and clearly showed the decorated dislocations running between etch pits, thus establishing the correspondence between pits and surface-emergent dislocations. Fig. 6.5 is a sketch of one of his results[4]. Dislocations can be made visible in many transparent crystals, for example by the precipitation of silver in silver halides[5] or silver decoration of potassium chloride[6], and numerous such photographs will be found in any book on dislocations[1, 2, 6, 7].

Because of their importance to the mechanical strength of materials most of the dislocation work has been based on metals, but for our purposes we must consider in more detail dislocations in non-metals, in particular ionic solids and covalent solids, since these include the crystalline insulators and semiconductors.

6.3 Dislocations in ionic solids

The free electron cloud smooths out electrical disturbances produced by discontinuities in metal crystal lattices. However, in ionic lattices this is not so, and wherever we have a dislocation we have a local charge disturbance. In producing an edge dislocation (Fig. 6.1) by shearing an ionic crystal along the glide plane AB, we are faced with the need to match up the ionic charges across the glide plane.

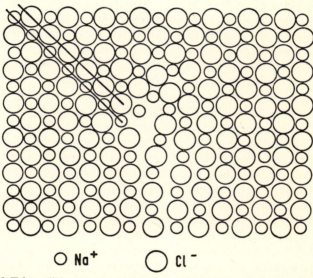

O Na⁺   O Cl⁻

Fig. 6.6 Edge dislocation produced in NaCl by inserting two half planes into the lattice.

This leads to dislocation configurations such as shown in Fig. 6.6, where we have inserted two half planes of ions into the lattice and produced an obvious disturbance of the local charge near the dislocation. The plane we have shown is the (010) plane in the NaCl lattice, and slip has been produced on the (10$\bar{1}$) plane. The configuration is not symmetrical for positive and negative charges, and the sign of the charge associated with these dislocations will alternate with each lattice plane as we move along the dislocation line in the

$<010>$ direction. A simple electrostatic calculation[8] shows that the effective charge at the point of emergence is $\pm\frac{1}{4}e$. The charge produces an additional force of interaction with charged point defects of opposite sign, encouraging them to cluster along the dislocations.

A jog, or indeed any kink, in a dislocation in an ionic crystal is also influenced by the additional conditions imposed by the need for local charge neutrality. A simple one-atom step jog in a pure edge dislocation has an effective charge of $\pm\frac{1}{2}e$ and so cannot be neutralized by point defects or impurities. If the jog step is two atoms high, the resultant superjog is neutral and the electrostatic force may dominate the repulsive elastic force. Unlike metals therefore, superjogs can be stable in ionic crystals. In general the charge on the jog is $\frac{1}{2}q$ where q is the ionic charge. Therefore jogs in divalent ionic crystals, like MgO, will have an effective charge of $\pm e$, and simple jogs can be neutralized by holes, electrons or charged point defects. There is then a reduced advantage in forming superjogs and a lower probability of the jog remaining charged. Trivalent atoms produce $\pm\frac{3}{2}e$ jog charges, and so on for higher valency atoms.

The geometry of screw dislocations is more complex, but the result is similar in that screw dislocations[2], like edge dislocations[9]. can have a net charge and therefore be influenced by external electric fields, produce image forces as they approach surfaces and interact coulombically. Charge can be transported by the motion of dislocations and jogs, but there is a difference in that jogs in edges move freely whereas jogs in screws do not.

The interaction between dislocations and point defects has been reviewed[10], and we will not pursue the subject in detail here, but we must remark that in ionic solids the most important interactions are electrostatic and the coulombic interaction energy between a charged jog and a point defect can be as large as 1 eV at interatomic distances[1]. The alternate positive and negative charges on a $<110>$ edge dislocation can produce short range forces of the same order between vacancies[11] or divalent impurities[1] and the dislocation core. The alternation of charges means that at distances much larger than atomic distances the coulombic forces average out.

6.4 Dislocations in the diamond lattice

The diamond lattice is adopted by elements which have covalent tetragonal bonding, such as silicon, germanium, grey tin and carbon in the form of diamond. The diamond lattice is actually two parallel interpenetrating face-centred cubic lattices, displaced by a quarter of a cube diagonal ($\frac{1}{4}$, $\frac{1}{4}$, $\frac{1}{4}$) with respect to each other. The structure is not as simple as a face-centred cubic lattice, because the environment of atoms in the two lattices is different. It can, however, be described as face-centred cubic with a basis of two atoms per lattice point.

Whereas metals arrange their structure so as to minimize their volume, under the influence of the free electrons, and ionic solids arrange for local charge neutrality, the diamond structure arises from the comparatively rigid tetrahedral covalent bonds. This makes them highly resistant to shear and brittle at room temperatures, but their ductility increases with temperature, particularly above 60% of their melting temperature. This ductile temperature is about 750°C

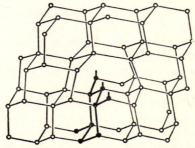

Fig. 6.7 A 60° dislocation along <110> in the diamond structure. The glide plane is (111). The extra half plane is shown in heavier lines.

for Si and 450°C for Ge. An advantage of this high resistance to shear is that crystals with very low dislocation densities (a few compared with the normal 10^6 cm^{-2} for other materials) are easy to grow. The rigid bonding leads to very localized dislocations[12] which occur in two sets corresponding to the two interpenetrating face-centred cubic lattices. One such dislocation is the 60° dislocation which predominates in deformed Ge (shown in Fig. 6.7) and leads to a series of unpaired valence bonds which are called dangling bonds. Those who wish to pursue the geometrical complexity of dislocations and jogs and their motions in the diamond lattice are referred elsewhere[2, 13, 14]. We will concentrate on the implications of the dangling bonds for defects and charge carriers. As at a free surface, the dangling bonds may affect the carrier concentration either by accepting an electron to complete the eight-electron configuration or by donating the remaining electron to the conduction band. Dislocations affect the electrical behaviour of n-type germanium much more than p-type, decreasing the concentration of carriers and ultimately changing n-type Ge to intrinsic Ge to p-type Ge[15]. It has hence been assumed[16] that the dislocations act only as acceptors and produce energy levels below the Fermi level E_d, but some more recent observations[17] have shown donor action by dislocations in germanium. The dislocation will become negatively charged if the bonds accept electrons, and be more likely to scatter electrons, and the conductivity of the material is therefore reduced. Not every bond site can be occupied, even at low temperatures, because of electrostatic repulsion between the electrons on adjacent bonds. For Ge with $a = 0.4$ nm, $E_F = 0.77$ eV, $E_d = 0.55$ eV and $N_D - N_A = 10^{15}$ cm^{-3}, N_D and N_A being the concentrations of donors and acceptors, Read[18] calculates an occupancy of only 0.11, that is one electron per 3.6 nm of dislocation at 0 K. The levels due to dislocations, which are below the Fermi level, do not obey Fermi statistics, which is not really surprising as the sites are highly ordered and not randomly distributed as are other point defects we have discussed.

There will in addition be an orientation effect. Charge carriers will be scattered if they attempt to flow through a forest of dislocations at right angles to them. The carrier mobility will be correspondingly decreased. There will be little effect on carriers moving parallel to the dislocations. Krauchenko[19] has measured this effect in n-type InSb with oriented dislocations produced by bending the crystal. The electron mobility was greatly decreased by the dislocations and varied with temperature both parallel and perpendicular to the dislocations in a manner consistent with a variation of point charged centres at the dislocations.

A general expression[20] which takes into account the effect on charge carrier mobility of vibration scattering, impurity scattering and dislocation scattering is

$$\frac{1}{\mu} = A_v T^{\frac{3}{2}} + A_i T^{-\frac{3}{2}} + A_d T^{-1} \qquad (6.1)$$

where μ is mobility and T is temperature, A_v, A_i and A_d are constants.

6.5 Internal friction and anelastic loss

If we stress a crystal by mechanical or electrical means, its response will be in terms of an elastic strain or an electrical polarization. The magnitude of the response is a characteristic of the material and for non-isotropic substances will be a function of orientation. For a perfect crystal and small stresses we have elastic behaviour, and the strain will reach its full value almost instantaneously and, when the stress is removed, return to its original value as quickly. A typical result of such an experiment on real materials is shown in Fig. 6.8. There is a delayed response to the application and removal of the stress. If the saturation strain is proportional to the stress and returns to zero when the field is removed, the behaviour is called anelastic relaxation[21]. The value of ε/σ, the strain per unit stress,

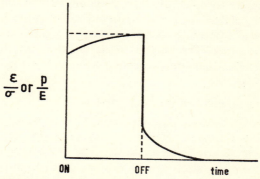

Fig. 6.8 The time variation of the strain per unit stress ε/σ or P/E the polarization per unit electric field under mechanical or electrical stress.

immediately after the mechanical stress is applied is called the compliance S and the additional amount to reach saturation is σS, the relaxation of the compliance. The electrical equivalents are the susceptibility or polarization per unit electrical field $k = P_0/E_0$ and the relaxation of the susceptibility σk. We will discuss the mechanical case here, and in the next section deal with the electrical behaviour under dielectric loss.

The growth and decay of the curves of Fig. 6.8 can be described by[22]

$$\frac{\varepsilon(t)}{\sigma} = S + \sigma S \left[1 - \exp \left(\frac{t}{\tau} \right) \right] \qquad (6.2)$$

where S is a characteristic of the perfect lattice and σS and τ are a function of the defects present. Under the application of the stress the defects rearrange themselves into new equilibrium positions, and the difference between the stressed and unstressed configuration determines σS. On the other hand τ is a measure of the speed with which the rearrangement occurs and is hence a characteristic of the kinetics of the rearrangement process.

In an ordered crystal the symmetry of the defects will determine whether relaxation occurs and which crystal parameters are involved. A detailed group theoretical analysis for all possible defect symmetries in all of the 32 crystal classes has been made by Nowick and Heller[22], who deduce selection rules for the coefficients of elastic compliance or electric susceptibility which relax in specific cases. The measurement of relaxation strengths and defect-induced changes in lattice parameter together with the selection rules enable the components of the elastic or electric dipoles to be determined. These are characteristic of a particular defect and help to identify and elucidate the defect structure. The analysis neglects defect reactions or the presence of more than one type of defect and is for low defect concentrations.

6.6 The measurement of internal friction

A cyclic stress is usually applied by a low-frequency mechanical system such as a torsion pendulum or by passing a high-frequency acoustic wave through the material. The perfect crystal would respond to the cyclic stress $\sigma = \sigma_0 \exp (i\omega t)$ with an in-phase strain ε_1, but in the presence of a relaxation period due to defects, the response will have a component ε_2 which is 90° out of phase with the applied stress where

$$\varepsilon = (\varepsilon_1 + i\varepsilon_2) \exp (i\omega t) \tag{6.3}$$

and ε_2 obeys the Debye equation[23]

$$\frac{\varepsilon_2(\omega)}{\sigma_0} = \sigma S \times \frac{\omega\tau}{1 + \omega^2\tau^2}. \tag{6.4}$$

A plot of log $\omega\tau$ versus $\varepsilon_2(\omega)/\sigma_0$ gives a symmetrical Debye peak centred at $\omega\tau = 1$ from which we get the value of τ. From equation (6.4) we see the other relaxation parameter σS is twice the peak height.

The methods of internal friction measurement are particularly sensitive for metals, but they have been applied to numerous defects in insulators, for example the O_2^- centre[24] in alkali halides and the A centre in irradiated silicon[24].

In general the analysis is complicated by the presence of more than one defect contributing to the relaxation processes and equation (6.4) becomes

$$\frac{\varepsilon_2(\omega)}{\sigma_0} = \sum_n \sigma S_n \frac{\omega\tau_n}{1 + \omega^2\tau_n^2} \tag{6.5}$$

which gives a relaxation spectrum with a set of relaxation times τ_n. For most point defects the temperature dependence of the relaxation times τ_n follows an Arrhenius equation

$$\frac{1}{\tau} = \left(\frac{1}{\tau_0}\right) \exp\left(-\frac{U}{kT}\right) \qquad (6.6)$$

where U is a migration energy.

One method of determining the relaxation times experimentally is to set the specimen into resonance using a transducer driven by a variable alternating current supply. The frequency of the resonance peak maximum ω_r is directly determined and the loss tangent $\tan \delta$ comes from the width $\Delta\omega$ of the peak at half maximum. From analogy with electrical resonant circuits

$$Q = \frac{\sqrt{3}\omega_r}{\Delta\omega} = \frac{1}{\tan \delta}. \qquad (6.7)$$

The relation between ω_r and the real part of the compliance S is[25]

$$\omega_r^2 = \frac{\gamma}{S_{real}} \qquad (6.8)$$

where γ is a function of the particular vibrating system used. The loss tangent gives us the imaginary part of the compliance from

$$\tan \delta = \frac{S_{imag}}{S_{real}}. \qquad (6.9)$$

In practice it is frequently more convenient to take advantage of equation (6.6) and vary τ by changing the temperature to plot out the Debye peak[26]. Alkali halides doped with aliovalent (differing in charge from the host ions) impurities have been extensively studied by a number of methods, including internal friction. The reason for the popularity of these systems is that the impurity associates with a nearby vacancy or interstitial ion to preserve local charge neutrality. The resultant dipole or defect complex[27] is fairly strongly bound and may be studied by relaxation methods and the ionic motions elucidated.

The decay of vibrations rather than a vibration driven at resonance is also used. The logarithmic decrement D of the decay is related to the loss angle and the Q of the process by

$$\frac{1}{Q} = \tan \delta = \log \frac{D}{\pi}. \qquad (6.10)$$

Observations[26] on NaCl doped with 200 p.p.m. $CaCl_2$ show that in the $<111>$ direction the results fit a single Debye peak with an activation energy of 0·7 eV. A multiple peak in the <100> direction was resolved into three Debye peaks, indicating that next nearest neighbours as well as nearest neighbours are involved in the inelastic relaxation and enabling the particular relaxations to be identified.

NaF in CaF_2 is an example of a system which has been studied[28] using both mechanically and electrically induced relaxation. We will look at it in some detail as an illustration of how the methods of analysis are applied. The sodium is associated with a neighbouring fluorine vacancy in CaF_2, and the structure of CaF_2 with and without a sodium ion is shown in Fig. 6.9. Internal friction measurements

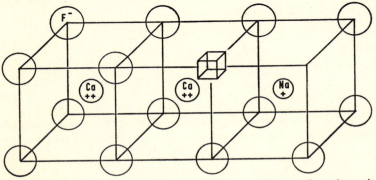

Fig. 6.9 The association of a fluorine vacancy with a sodium impurity ion in CaF_2.

were taken by driving the specimen rod with a loudspeaker and wide band oscillator and detecting the sample vibration amplitude with a gramophone pickup. The amplitude decay was observed on a cathode ray oscilloscope. Inelastic loss was measured at a number of temperatures and frequencies, changing the resonant frequencies by changing the length of the sample. The internal friction peak was independent of sodium concentration and is shown in Fig. 6.10 together with a Debye peak calculated for an activation energy of 0·53 eV. This compares well with 0·55 eV found for the diffusion activation energy[29] of a fluoride vacancy in CaF_2. From the frequency of the loss peak at each temperature the pre-exponential factor in equation (6.6) was measured as $3·29 \times 10^{-15}$ seconds. The mechanical relaxation measurements thus give

$$\tau = (3·29 \times 10^{-15}) \exp\left(\frac{0·53}{kT}\right). \tag{6.11}$$

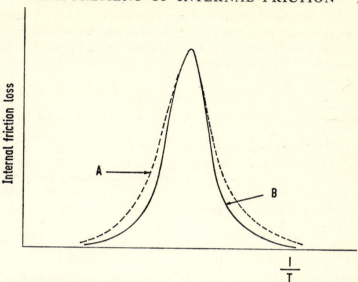

Fig. 6.10 A comparison of the experimental A and calculated B Debye
peak for internal friction loss measurements in CaF_2 : Na. The
calculated curve is for an activation energy of 0·53 eV.

The mechanism of relaxation is the movement of the fluorine vacancy
from one site to another, the sodium atom remaining fixed. If the
jump is to one of the eight nearest neighbour sites[30], samples cut
with $<100>$ axis should not show a loss peak whereas the $<111>$ the

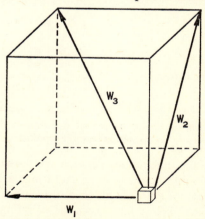

Fig. 6.11 The three possible directions for the jump of a fluorine vacancy
adjacent to a substitutional Na in CaF_2.

<110> and the polycrystalline samples should show a peak. This was in fact observed. There are three possible jumps available in such a system (Fig. 6.11) with ion jump frequencies W_1, W_2 and W_3, and an analysis of the normal modes[30, 31] gives the relaxation modes for mechanical stress as

$$\lambda_{\text{mech}} = 4(W_1 + W_2). \tag{6.12}$$

We will compare this with the electrical stress case later (§ 6.8).

6.7 Dislocation motion and internal friction

In alkali halides much of the internal friction is due to the motion of dislocations, which can be thought of as vibrating like strings between fixed ends called pinning points. Vibration within a viscous medium causes an absorption of energy, and this energy loss can be

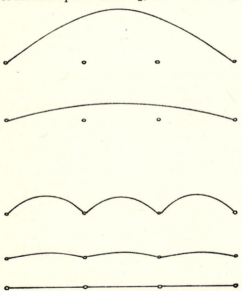

Fig. 6.12 The vibrating string model of dislocation damping showing breakaway from the weaker pinning points at high strain.

used to measure the interactions between the dislocations and their surroundings. Pinning points may be any irregularity in the lattice such as a vacancy, an impurity atom, a dislocation jog or an intersecting dislocation. Their ability to pin the vibrating dislocation will vary with their nature, and as the strain level is increased the dislocation will break away from the weaker pins first (Fig. 6.12), thus changing the vibrating loop length and the absorption characteristics of the sample. One can thus obtain information on the distribution and nature of the pinning points and hence on the defects involved.

The viscous damping has been attributed to interactions with phonons in the Granato Lücke[32] theory, which was originally developed for metals. Assuming a random distribution of loop lengths and ignoring breakaway effects the theory gives a decrement (the energy lost per cycle divided by twice the maximum energy stored per cycle) which is proportional to the fourth power of the loop length. With the onset of breakaway the loop length increases and hence the decrement is greatly increased. The theory was later extended to take account of such additional effects as thermal fluctuations[33]. The theory has been applied with success to a number of observations, for example deformed NaCl crystals[34] and up to strain amplitudes of 5×10^{-6}[35]. There is some evidence that pinning points may move along the dislocation lines before breakaway[36], which does not seem unreasonable.

Any process which reduces loop length will have the opposite effect to high strain breakaway. It will rapidly reduce the damping. For example, the damping of NaCl at 85 kHz can be reduced almost to zero by irradiation with X-rays, and a number of studies of the pinning effects of X-ray irradiation on alkali halides have been made[37, 38]. It was found[37] that pinning due to irradiation at low temperatures could be unpinned by light, which suggested the model of a pinning complex consisting of a jog at a halide ion with an F centre one-atom spacing away below the glide plane. Ionization of the F centre by light enables it to recombine with the halide ion removing the pinning. Internal friction measurements have been made on alkali halides irradiated by gamma rays[39, 40] and by neutrons[41], and the observations have usually been interpreted on the basis of the Granato Lücke theory.

We have seen in § 6.2 that, unlike metals, ionic solids have charged dislocations[9], and this forms the basis of an alternative mechanism of internal friction in alkali halides proposed by Robinson and Birnbaum[42]. They suggest that it is the coulombic interaction of charged dislocations with their surrounding compensating charge clouds that provides the damping mechanism. Both theories predict a damping peak at about 10 MHz with the Robinson–Birnbaum theory giving the broader peak. The phonon damping theory gives a damping constant which is frequency-independent, whereas the charge cloud mechanism leads to a frequency-dependent damping constant. In a recent paper Robinson[43] has analysed his results on KCl at 40, 120 and 200 kHz, those of Suzuki et al.[44] on LiF and KCl from 5 to 100 MHz and those of Mitchell[45] on LiF from 20 to 500 MHz. He concludes that, under irradiation, defects are more easily formed at dislocations than in the perfect crystal regions and that the results fit the charged cloud theory better than the phonon damping theory. In particular, the damping predicted by the phonon damping theory is too high in the kHz region.

6.8 Dielectric absorption

The behaviour of a dielectric under an electric field is similar to its behaviour under a stress field described in the previous section. In the presence of defects which can relax under the applied field

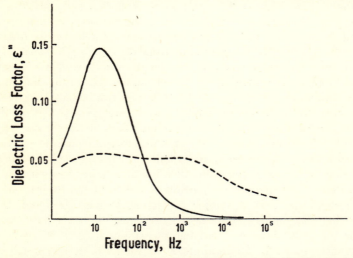

Fig. 6.13 Dielectric loss as a function of frequency for NaCl with $3 \cdot 6 \times 10^{-4}$ mole fraction of Ca^{++}. The narrow peak is obtained by measuring at 20°C immediately after quenching from 130°C. The broad double peak is obtained after storage at 20°C and the second peak is due to aggregation by diffusion.

a curve such as Fig. 6.8 will be obtained. Where a cyclic field of frequency ω is applied and a single relaxation period τ is operating, the dielectric loss factor ε'' will follow a Debye curve given by

$\varepsilon'' = B\dfrac{\omega\tau}{1 + \omega^2\tau^2}$, and the temperature variation given by equation

(6.6). An example of such a curve[46] is shown in Fig. 6.13 for NaCl doped with Ca^{++} immediately after quenching from 130°C to 20°C, in which we obtain a single peak indicating one set of dipoles producing the dielectric absorption. If kept at 20°C for some time the centres aggregate and produce a complex with a shorter relaxation time which produces the peak at a higher frequency. Dryden and his associates have studied the aggregation of divalent cationic impurities in alkali halides using this dielectric absorption method[27,47,48]. They suggest that the aggregate is a cluster of three vacancy impurity dipoles because the reaction approaches third order kinetics. Aggregation via an intermediate stage of two dipoles will also produce third order kinetics, if one makes the appropriate choice of rate constants. (The structure of the trimer is shown in Fig. 9.15 which is a section through a (111) plane of a LiF structure with its hexagonally arranged positive ions; this trimer is a hexagon composed of 3 M^{++} ions alternating with vacancies, each such dipole preserving local charge neutrality and lying in the (111) plane.) At room temperature most divalent cationic impurities are coupled with vacancies in this way. Further aggregation occurs by the addition of two dipoles at a time, again following third order kinetics[49]. The limiting process in trimer formation is diffusion of the dipoles through the lattice, as shown by the activation energy obtained from the kinetics of trimer formation. There is a small difference in the activation energies for aggregation and for dipole diffusion, and this is the association energy of the trimers which for Mg^{++} in LiF is found[49] to be 0·3 eV per dipole or 0·9 eV for the trimer.

We will now return to the NaF in CaF_2 system whose mechanical behaviour we have already discussed. Dielectric loss measurements on a capacitance bridge as a function of temperature and frequency[28] give both the activation energy and the pre-exponential factor as

$$\tau = (7 \cdot 25 \times 10^{-15}) \exp\left(\frac{0 \cdot 53}{kT}\right). \qquad (6.13)$$

Comparison with equation (6.11) shows the same activation energy of 0·53 eV and agreement in the pre-exponential factor within a factor of two. In the electrical case an analysis of the normal modes[30, 31] indicates that the body diagonal jump W_3 (Fig. 6.15) is allowed by comparison with equation (6.12) for the mechanical case

$$\lambda_{elec} = 2W_1 + 4W_2 + 2W_3. \qquad (6.14)$$

Thus a comparison between equations (6.12) and (6.15) shows that the W_1 jump, in the <100> direction (Fig. 6.11) is twice as likely to occur in the mechanically stressed case as in the case of electrical stress. Measurements of the differences in relaxation time for different crystallographic directions and for different methods of stressing can hence yield useful information on the atomic defect motions which occur.

A number of refinements are necessary in relating the basic relaxation measurements to atomic relaxation times[28]. Young's modulus may differ slightly at frequencies above and below the mechanical loss peak. Similar changes may occur in the dielectric constant on either side of the dielectric loss peak and internal fields acting on the dipole must be corrected for[28].

The dielectric relaxation processes in lithium sodium and potassium halides have been reviewed by Dryden and Meakins[50], who give tables of data including frequency factors and activation energies obtained by dielectric absorption, d.c. conductivity, diffusion and nuclear magnetic resonance.

6.9 Ionic conductivity

Ionic conductivity measurements have the advantage of being experimentally simple. For example, graphite electrodes can be applied to alkali halide specimens of convenient size and standard d.c. or low-frequency a.c. measurements made with straightforward electronic measuring devices[51]. The difficulty comes in analysing the results. Even if we restrict ourselves only to the predominant Schottky defects and perfectly pure crystals, we need at least six parameters to discuss the conductivity in terms of point defects. We need the enthalpy and entropy associated with Schottky defects (equal numbers of positive and negative ion vacancies), and with the motions of both positive and negative ion vacancies. In any real case we must also know the parameters for association between vacancies and aliovalent impurities and we must consider coulombic interactions and the contribution of more complicated types of defects.

Fortunately measurements over a range of temperatures enable the separation of different contributions to be made. At high temperatures intrinsic or thermally made defects predominate. At low temperatures intrinsic defects or defects due mainly to aliovalent impurities play the major role. Another region can frequently be separated in which impurity ions and adjacent vacancies associate and move together. Even with this simplification the analysis is not easy, but this has not prevented the accumulation of a large body of experimental data which has been reviewed by Lidiard[52] and Süptitz and Teltow[53].

The problem of fitting the conductivity data obtained by experiment to the theories which involve so many parameters is an obvious job for a computer, and it is perhaps a little surprising that the first such attempt was not made till 1966 when Beaumont and Jacobs[54] fitted their data on the conductivity of KCl over a wide temperature range. They assumed that defects were non-interacting unless two oppositely charged defects became nearest neighbours, after which they were considered as an uncharged complex. A number of other computer-aided analyses have since been made, for example on KCl[55] pure and doped with strontium chloride or potassium carbonate, and NaF[51] pure and doped with calcium fluoride. Both of these papers give tables of the enthalpy and entropy parameters deduced from computed least squares fit which selects the best value of the parameters by minimizing S where

$$S = \Sigma[(\log \sigma T)_{expt} - (\log \sigma T)_{calc}]^2 \qquad (6.15)$$

and σ is the conductivity and T the temperature.

In terms of the charge carrying species of concentration n the conductivity is

$$\sigma = \Sigma n q \mu \qquad (6.16)$$

where q and μ are charge and mobility and [52]

$$\mu = 4a^2 e \frac{W}{kT} \qquad (6.17)$$

where a is the cation-anion separation, e the electronic charge and

$$W = \nu \exp\left(-\frac{\Delta g}{kT}\right)$$

where ν is the vibration frequency of the ion on its original site and Δg is the free energy for motion. In NaF both anion and cation vacancies contribute to conduction[51], so assuming both vibration frequencies to be the same we have

$$\sigma T = \frac{4Ne^2a^2\nu}{k}\left[x_1 \exp\left(-\frac{\Delta g_m}{kT}\right) + x_2 \exp\left(-\frac{\Delta g'_m}{kT}\right)\right] \qquad (6.18)$$

for a fraction x_1 of cation vacancies and x_2 of anion vacancies where $\Delta g'_m$ is the free energy of motion of the cation vacancies. The factor 4 arises from the number of possible jumps available to the vacancy.

Other effects which this simple theory overlooks are coulombic interactions, a drag-factor due to the Debye–Hückel charge cloud and temperature variation of the quantities such as dielectric constant and vibration frequency. Some of these have been taken into account by more complete theoretical treatments of which recent reviews are available[56, 57], but a number of as yet unexplained discrepancies remain which cannot be due only to electronic conduction.

6.10 Lattice parameter, length and density changes

As the temperature of a crystal is raised the number of vacant lattice sites increases. If no atoms are lost, the number of unit cells must increase, perhaps by adding extra surface layers. Thus the density decreases as the macroscopic size of the specimen increases. The macroscopic size increases by a larger fraction than does the lattice parameter as the temperature is increased.

An accurate measurement of both lattice parameter and a dimension, or the density, of the crystal will hence reveal the number of vacancies formed as the temperature increases. Conversely if the predominant point defects formed by temperature change or irradiation are interstitials, then the crystal will lose unit cells as atoms take up interstitial positions. If both vacancies and interstitials are formed in equal numbers no change will occur. X-ray measurements of fractional lattice parameter changes $\Delta a/a$ have been carried out for a number of metals, together with the fractional length changes, and one such result is shown in Fig. 6.14 for aluminium[58]. For Schottky defects the changes in length and lattice parameter are related to the vacancy concentration C_v by

$$\frac{1}{3}C_v = \frac{\Delta l}{l} - \frac{\Delta a}{a}.$$

In pure alkali halides the predominant defects produced by thermal treatments are Schottky defects (i.e. interstitials diffuse to the surface) which lead to a density decrease. In silver halides the predominant defects are Frenkel, vacancy plus interstitial with no density change.

The introduction of aliovalent impurities, say Ca^{++} in KCl, also lowers the density in proportion to the amount of Ca^{++} present. The demand for local charge neutrality leads to a close association between the Ca^{++} and an adjacent positive ion vacancy, the number of vacancies increasing with the number of Ca^{++} ions present. More vacancies mean lower density, and this is found to be so, despite the fact that the calcium ion is heavier than the potassium ion it replaces.

A number of experimental techniques have been used to measure

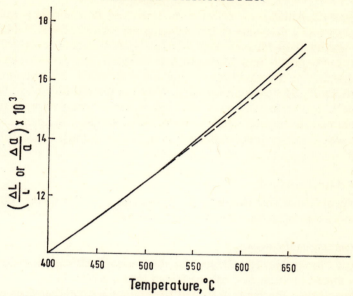

Fig. 6.14 The fractional length change $\Delta l/l$ and the fractional change in lattice parameter $\Delta a/a$ for aluminium as a function of temperature.

the small density and length changes associated with defect creation such as photoelasticity[59], density changes [60, 61] or various sensitive transducers to measure the small length changes involved[62]. Interpretation is complicated in the case of irradiation experiments by the production of other centres along with the simple vacancies we have discussed and irradiation damage also produces Frenkel defects. Careful control of irradiation and temperature is needed in this kind of experiment, and when this is done measurements of the volume changes associated with specific defects can be obtained; for example, in low temperature X-ray-irradiated KCl it is found[62] that the volume change for a Frenkel pair is

$$\frac{\Delta V}{V} = 1 \cdot 3 \pm 0 \cdot 2.$$

Radiation-induced density changes as low as a few parts in 10^6 can be determined by a flotation method in which normal crystals and irradiated crystals are immersed in a tube with a temperature gradient. Such measurements on LiF, NaF, NaCl and KCl after gamma-ray irradiation in a reactor[61] have been used to study the effects of impurities on defect formation. The measurements indicated that at a dose rate of 200 Röntgen per second ($\Delta\rho/\rho$) for LiF was greater than for NaF and that for NaCl was greater than for KCl. O^{--}, OH^{--} and Mg^{++} in LiF increased the rate of defect production, as did Ca^{++}, Sr^{++} and Ba^{++} in NaCl. The measurements also indicate the importance of back reactions during irradiation[63], and these are obviously influenced by the purity of the crystal and the irradiating conditions. The methods, in spite of these difficulties, can produce reliable results and are fairly extensively used.

6.11 Other methods

We will close this chapter with a mention of a few of the other methods that have been used to study defects.

Ionic thermal currents

Any dipoles in a crystal, such as those associated with aliovalent ions linked to vacancies, can be frozen into a crystal in a polarized state and then thermally activated to produce a small displacement current[64]. Because each type of dipole has a characteristic energy the technique is useful in studying each type of impurity dipole in turn. The experimental procedure is to polarize the dipoles at a temperature T_p, which is sufficiently high that the dipoles can rotate, because of the electric field, but the dipoles do not have enough energy to become thermally disordered. The sample is then cooled and the electric field E_p is removed. On warming the crystal at a steady rate $\beta = (\mathrm{d}T/\mathrm{d}t)$ one observes a polarization current which can be shown to be given by

$$i(T) = \frac{Np^2\alpha E_p}{kT_p}\left[\frac{\exp\left(-\int_0^T (\beta\tau_0 \exp(E/kT))^{-1}\,\mathrm{d}T\right)}{\tau_0 \exp(E/kT)}\right]$$

where p is the dipole moment, τ is the relaxation time, α is a geometrical factor and E is the energy of relaxation of the defect.

In addition to providing τ_0 and E the area under each current peak is proportional to the dipole moment per unit volume for each type of defect. The technique is fairly sensitive, and a dipole concentration of the order of 10^{-7} mole fraction can be detected.

All electrical measurements of insulating materials have inherent problems of contacts and space charge regions, but the technique of ionic thermocurrents is of value in studying impurity ion complexes.

Time-dependent electrical polarization

An alternative to freezing dipoles in and releasing them by warming is suddenly to change the electric field applied to the crystal. The polarization change produces a transient current which decays exponentially with one or more time constants which can be related to specific defect processes. The method is about an order of magnitude less sensitive than ionic thermocurrent methods.

Additional heat capacity

Heat capacity measurements at constant pressure show an anomalous increase at high temperatures as the solid approaches its melting point, which cannot be explained in terms of the perfect lattice. The extra capacity comes from the energy absorbed by the crystal in creating additional defects. Some observations on the heat capacity of silver bromide[65] are shown in Fig. 6.15.

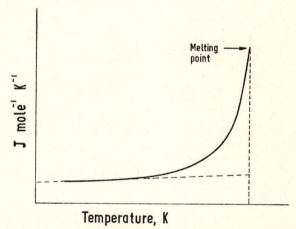

Fig. 6.15 The heat capacity of silver bromide increases rapidly at high temperatures because of the energy required to form defects in this region.

The release of stored energy

Crystals irradiated at one temperature will retain defects which cannot anneal at that temperature. When heated the defects will anneal when the temperature is sufficient to provide the activation energy for relaxation, and the energy released can be measured by calorimetry[66]. An irradiated and an identical unirradiated crystal are heated together in the calorimeter and a differential thermocouple determines any temperature rise in the damaged sample due to energy release with a sensitivity of about 40 joules kg^{-1}. Such measurements can usefully be combined with thermoluminescence and thermally stimulated current observations.

Stress strain curves

The elastic properties of a solid depend not only on dislocations but on the nature and distribution of point defects. The application of a compressive stress applied at a uniform rate and sensitive measurement of the resultant strain has been used[49] to study the effect on hardness of aliovalent impurities such as Mn^{++}, Sr^{++}, Ba^{++} and Mg^{++} in NaCl, KCl and LiF. The increase in critical shear stress was found to be proportional to the two-thirds power of the aliovalent ion-vacancy dipoles. A large increase in hardness was found as trimers in the LiF : Mg^{++} system aggregated into larger aggregates.

Chapter 6 References

[1] Friedel, J., *Dislocations* (Pergamon) 1964.

[2] Hirth, J. P., and Lothe, J., *Theory of Dislocations* (McGraw Hill) 1968.

[3] Hirsch, P. B., *Phil. Mag.* **7**, 67, 1962.

[4] Dash, W. C., in *Dislocations and Mechanical Properties of Crystals* edited by J. C. Fisher, W. G. Johnston, R. Thomson and T. Vreeland (Wiley) 1957.

[5] Jones, D. A. and Mitchell, J. W., *Phil. Mag.* **3**, 1, 1958.

[6] Amelinckx, S., *Acta Metall.* **6**, 34, 1958; *Solid State Physics* **13**, 147, 1962; Supp. **6**, *Solid State Physics*, 1964.

[7] Newkirk, J. B. and Wernick, J. H., *Direct Observation of Imperfections in Crystals* (Interscience) 1962.

[8] Seitz, F., *Phys. Rev.* **80**, 239, 1950.

[9] Eshelby, J. D., Newely, C. W. A., Pratt, P. L. and Lidiard, A. B., *Phil. Mag.* **3**, 75, 1958.

[10] Bullough, R. and Newman, R. C., *Rep. Prog. Phys.* **33**, 101, 1970.

[11] Bassini, F. and Thomson, R., *Phys. Rev.* **102**, 1264, 1956.

[12] Hornstra, J., *J. Phys. Chem. Solids* **5**, 129, 1958.

[13] van Bueren, H. G., *Imperfections in Crystals* (North Holland) 1960.

[14] Alexander H. and Haasen, P., *Solid State Physics* **22**, 27, 1968.

[15] Tweet, A. G., *Phys. Rev.* **99**, 1245, 1955.

[16] Shockley, W., *Electrons and Holes in Semiconductors* (van Nostrand) 1950.

[17] Schröter, W., *Phys. Status Solidi*, **21**, 211, 1967.

[18] Read, W. T., *Phil. Mag.* **45**, 775, 1954, **46**, 111, 1955.

[19] Krauchenko, A. F., *Fiz. Tekh. Poluprovodnikov* **3**, 1346, 1969.

[20] Dexter, D. L., and Seitz, F., *Phys. Rev.* **86**, 964, 1952.

[21] Zener, C., *Elasticity and Anelasticity in Metals* (University of Chicago Press) 1948.

[22] Nowick, A. S. and Heller, W. R., *Advances in Physics* **14**, 102, 1965.

[23] Frölich, H., *Theory of Dielectrics* (Oxford) 1949.

[24] Nowick, A. S. and Heller, W. R., *Advances in Physics* **12**, 251, 1963.

[25] Nowick, A. S. and Berry, B. S., *Anelastic Relaxation in Crystalline Solids* (Academic Press) 1971.

[26] Dreyfus, R. W. and Laibowitz, R. B., *Phys. Rev.* **135**, A1413, 1964.

[27] Dryden, J. S., *J. Phys. Soc. Japan* Supplement III, **18**, 129, 1963.

[28] Johnson, H. B., Tolar, N. J., Miller, G. R. and Cutler, I. B., *J. Phys. Chem. Solids*, **30**, 31, 1969.

[29] Bontinck, W., *Physica* **24**, 650, 1958.

[30] Wachtman, J. B., *Phys. Rev.* **131**, 517, 1963.

[31] Chang, R., *J. Phys. Chem Solids*, **25**, 1081, 1964.

[32] Granato, A. V. and Lücke, K., *J. Appl. Phys.* **27**, 583, 1956.

[33] Granato, A. V., in *Dislocation Dynamics* edited by A. R. Rosenfield (McGraw-Hill) 1968.

[34] Bauer, C. L. and Gordon, R. B., *J. Appl. Phys.*, **31**, 945, 1960.

[35] Phillips, D. C. and Pratt, P. L., *Phil. Mag.* **21**, 217, 1970.

[36] Platkov, V. Ya., *Sov. Phys. Sol. State* **11**, 343, 1969.

[37] Bauer, C. L. and Gordon, R. B., *J. Appl. Phys.* **33**, 672, 1962.

[38] Gordon, R. B. and Nowick, A. S., *Acta. Metall.* **4**, 514, 1956.

[39] Truell, R., *J. Appl. Phys.* **30**, 1275, 1959.

[40] Silvertsen, J. M., *Acta. Metall.* **10**, 401.

[41] French, I. E. and Pollard, H. F., *J. Phys. C.* **5**; *Solid State Physics* **3**, 1866, 1970.

[42] Robinson, W. H. and Birnbaum, H. K., *J. Appl. Phys.* **37**, 3754, 1966.

[43] Robinson, W. H., *J. Mater. Sci.* 7, 115, 1972.

[44] Suzuki, T., Ikushima, A and Aoki, M., *Acta Metall.* **12**, 1231, 1964.

[45] Mitchell, O. M. M., *J. Appl. Phys.* **36**, 2083, 1965.

[46] Cook, J. S. and Dryden, J. S., *Aust. J. Phys.* **13**, 260, 1960.

[47] Cook, J. S. and Dryden, J. S., *Proc. Phys. Soc.* **80**, 479, 1962.

[48] Dryden, J. S. and Harvey, G. G., *J. Phys, C.* **2**, 603, 1969.

[49] Dryden, J. S., Setsu Morimoto, and Cook, J. S., *Phil. Mag.* **12**, 379, 1965.

[50] Dryden, J. S. and Meakins, R. J., *Disc. Faraday Soc.* **23**, 39, 1957.

[51] Bauer, C. F. and Whitmore, D. H., *Phys. Stat. Sol.* **37**, 585, 1970.

[52] Lidiard, A. B., *Handbuch der Physik* **20**, 246 (Springer-Verlag, Berlin) 1957.

[53] Süptitz P. and Teltow, J., *Phys. Stat. Sol.* **23**, 9, 1967.

[54] Beaumont, J. H. and Jacobs, P. W. M., *J. Chem. Phys.* **45**, 1496, 1966.

[55] Chandra, S. and Rolfe, J., *Canad. J. Phys.* **48,** 412, 1970.

[56] Tosi, M. P., Nat. Bureau. Std. Miscell. Publn. No. 287, 1, 1967.

[57] Barr, L. W. and Lidiard, A. B., A.E.R.E. Harwell Report TP356.

[58] Simmons, R. O. and Balluffi, R. W., *Phys. Rev.* **117**, 52, 1960.

[59] Merriam, M. F., Wiegand D. A. and Smoluchowski, R., *J. Phys. Chem. Sol.* **25**, 273, 1964.

[60] Bleckmann, A. and Thommen, K., *Z. Phys.* **191**, 160, 1966.

[61] Andreev, G. A. and Vasilev, G. Ya., *Sov. Phys. Sol. State* **10**, 1030, 1968.

[62] Balzer, R., Peisl, H. and Waidelich, W., *Phys. Stat. Sol.* **31**, K29, 1969.

[63] Sonder, E. and Sibley, W. A., *Phys. Rev.* **140**, A539, 1965.

[64] Bucci, C., Fieschi R. and Guidi, G., *Phys. Rev.* **148**, 816, 1966.

[65] Christy, R. W., and Lawson, A. W., *J. Chem, Phys.* **19**, 517, 1951.

[66] Bunch, J. M. and Pearlstein, E., *Phys. Rev.* **181**, 1290, 1969.

7

SPECIFIC DEFECT MODELS

7.1 Introduction

Without doubt the simplest and most positive identification of a defect comes from the resonance measurements discussed in Chapter 2. In particular, impurity centres are simple to identify because of the control of isotopic abundance and the total defect concentration.

For example, in rutile TiO_2, ESR and ENDOR spectra have been quoted for such impurities as Cr, Fe, Nb, Co, Mo and V. Unfortunately not all defects are suited for resonance measurements because they do not have unpaired electrons, or the spectra are too complex, different spectra overlap or the lines are quenched by the microwave power needed for ESR. In such cases the value of the other techniques becomes apparent.

For the examples which will now be quoted reference will be made to resonance data, but from the assumed model of the defect we will also predict the other properties that should be observed. It is then a question of personal judgement to state whether or not one could derive the same model from these facts alone.

It should also be remembered that resonance data are generally obtained for the ground electronic state and may not be of value in predicting properties of excited states, the interaction between defects, or the mechanisms of defect formation and migration.

7.2 Electron centres in alkali halides

The F centre must have received more attention than any other and it is well understood. It also has the distinction of having an entire book[1] devoted to its properties–or at least those known up to 1966!

Its importance is not in the depth with which this particular centre is known, but rather because it is a prototype for all defect studies. It is the simplest of defects, being a halogen vacancy which has trapped an electron (Fig. 7.1) to re-establish the electrostatic equilibrium of the lattice. Therefore one expects all insulators to show comparable defects where anions are displaced but charge equilibrium is

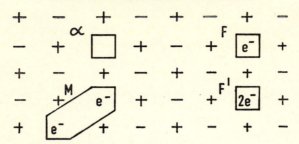

Fig. 7.1 The simple halogen vacancy defects in alkali halides:
  α[F⁺] —a halogen ion vacancy
  F   —a halogen ion vacancy plus a trapped electron
  F′[F⁻]—a halogen ion vacancy plus two trapped electrons
  M[F₂] —two adjacent F centres.

maintained. From a theoretical viewpoint the F centre is sufficiently simple that one can compute the energy levels and the lattice interactions of both the ground and excited states and test the calculation against experimental observations. It is also central to defect studies because there is the possibility of other charge states and the association of simple defects into more complex structures of lower symmetry.

To simplify the discussion of the F centre we shall first list the models of the defects which produce optical obsorption bands in alkali halides and are directly related to the F centre. Not all centres

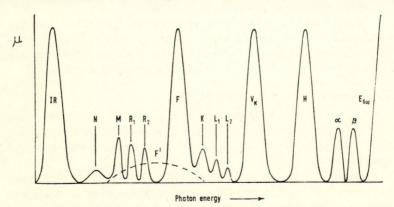

Fig. 7.2 The relative positions of the major optical absorption bands in alkali halide crystals. In perfect crystals there would only be the infra-red lattice absorption (IR) and the ultra-violet fundamental absorption edge (E_{GAP}).

are equally well established at this stage, but it would seem reasonable to assume the existence of simple defects even if they are not responsible for the optical absorption bands that one initially assigned to them. Such an approach has an excellent precedent in the 1946 and 1954 review papers by Seitz[2]. In his work several defect models were suggested and tentatively related to particular absorption bands. The proposed models then focused attention on these centres and predictions of their properties, so that as more techniques and results became available it was possible to make more positive identifications.

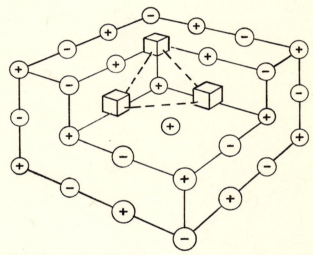

Fig. 7.3 A sectional view of three F centres arranged in a (111) plane to form an R[F₃] centre.

For simplicity an idealized set of absorption bands is shown in Fig. 7.2. Experimental curves show impurity bands, and it is not possible to present actual data showing all these features because the peaks overlap. The conditions of production and measurement also preclude the appearance of all these peaks simultaneously.

Currently accepted models for the halogen vacancy defects are shown in Figs. 7.1, 7.3 and 7.4, and the description of such centres is given by the following list in terms of the historical label used for the absorption band. A more logical set of labels for the defects has been suggested by Sonder and Sibley[3] in which the charge state of the defect, compared with the neutral lattice, is indicated by a superscript and the number of associated F centres by a suffix. The alkali

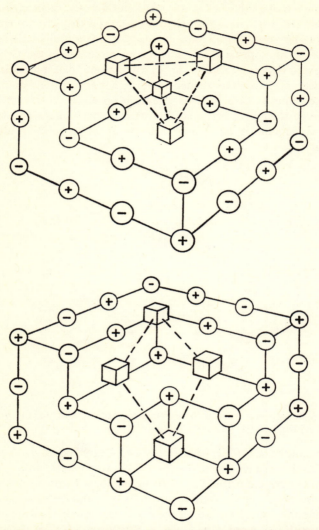

Fig. 7.4 Two possible arrangements of four F centres to form N centres. These arrays are either a tetrahedron or a parallelogram.

halide notation is probably too firmly entrenched to change at this stage, but in other materials such a notation is clear. However, one still has the problem of relating the models to particular absorption bands. Hole traps will be discussed later.

α band (F⁺)—a halogen ion vacancy which modifies electronic transitions of the lattice.

β band—a lattice excitation near an F centre.

F band (F)—a halogen ion vacancy that has trapped an electron.

K, L_1, L_2, L_3 bands—absorption produced by transitions of the F centre to higher states.

F_A, F_B bands—(sometimes labelled A and B)–absorption bands which arise from impurity alkali ions at one of the six neighbouring metal ions sites of an F centre. (There are two bands because of the reduced symmetry of the centre, and the dipole may lie along the axis through, or perpendicular to, the impurity ion.)

F' band (F⁻)—a halogen ion vacancy that has trapped two electrons.

M band (F_2)—a pair of adjacent F centres.

R bands (F_3)—a triangular array of three F centres. Although possible arrangements are a linear chain of three vacancies, an L-shaped set in the (100) plane or a triangular set in the (111) plane, the model currently believed is the latter (111) arrangement.

N bands (F_4)—four F centres linked together. Two possible configurations are a parallelogram of vacancies in the (111) plane which may produce the N_1 absorption and a tetrahedron of vacancies which may give the N_2 absorption.

M', R', N' bands (F_2^-, F_3^-, F_4^-)—by analogy with the F centre the M, R or N centres may trap an additional electron. Some absorption bands have been associated with these defect models and are discussed in the review article of F aggregate centres by Compton and Rabin[4].

The F centre

We shall now summarize the steps required to arrive at the model for this classic defect. The F centre is the dominant absorption band in all alkali halides and is produced by all types of treatment such as irradiation, electrolytic coloration, quenching and changes in stoichiometry. Therefore it is a simple intrinsic defect, and since the position of the absorption band is independent of the alkali metal used to colour the crystal by non-stoichiometry, the two obvious models are a halogen vacancy or a metal interstitial. We may also guess that the vacancy site would trap an electron, since this maintains the charge neutrality of the perfect lattice. In this instance the choice of models is made by considering the ESR and ENDOR signals, as described in Chapter 2. For a suitable choice of isotope (say K^{39} in KCl) the defect will exhibit a set of hyperfine energy levels from the nuclear interaction. If the model of a metal ion interstitial were correct the nuclear spin of $3/2$ and a K^{39} coupling constant of 0.007 cm^{-1} would produce a four-line ESR spectra spread over 249 gauss. This is not found.

The alternative model of an electron trapped in a halogen vacancy will have six equal interactions with metal neighbours, so we expect a 19-line spectra, or if these are not resolvable, a Gaussian envelope of half-width of 47 gauss as mentioned in Chapter 2. Not only is this found, but also detailed ENDOR studies reveal the more distant neighbour interactions.

Without the aid of resonance results an alternative approach to a defect such as this might have been to compute the energy levels for the two possible systems and compare the transition energies with the position of the optical absorption band.

From this correct model of the F centre we can predict some of the properties of the system. Very briefly these are:

(i) The electron is trapped in a potential well which is physically determined by the lattice spacing, so the transition energy for the F absorption should be a simple function of the lattice spacing. In this case the Mollwo-Ivey law is followed with $E \propto d^{-1.84}$.

(ii) The defect contains a trapped electron, so might give photoconductivity.

(iii) The defect symmetry can be tested by polarized light (as well as in ESR experiments).

(iv) The excited state relaxes and decays by a luminescence process which has a large Stokes shift. Therefore we find infra-red luminescence which can be tested for polarization to check on the symmetry of the excited state.

(v) The electron is not completely confined to one lattice site, so the

F centre is slightly positive compared with the lattice and can weakly attract a second electron. This will produce a broad low-energy absorption band, which will also show photo-conductivity and will be thermally unstable except at low temperatures. This is found as an F′ centre.

(vi) Alternatively the simple centre may associate with other centres, and by noting the relative changes in concentration it is possible to decide how many centres are involved. These are the M, R, N series of bands for two, three and four F centres.

(vii) Similarly a point defect in a perfect lattice may still be recognized even if the lattice is distorted in some way. There are many examples of this for F centres. For example, consider the effect of changing one of the neighbouring metal ions to an impurity alkali ion[5]. The symmetry of the centre is reduced as it will now have one axis through the impurity atom and transitions oriented along this axis will differ from those perpendicular to the axis. Therefore we expect two absorption bands (F_A, F_B) slightly displaced from the position of the original F centre. Such centres will behave differently under the action of light polarized along, or perpendicular to, the defect axis.

Changes in peak position and defect stability can also occur if the F centre experiences an electric field. This is the case for F centres in the vicinity of dislocations where the space charge around the dislocation results in a shift of the peak position[6]. More drastic perturbations are produced by association with multiple-charged impurities. The Z bands lie close to the position of the normal F band and result from F centres linked to divalent impurities.

All these possible forms of F centre will have different thermal stabilities, and even if the broad F band disguises their presence they can be detected as a series of annealing stages of nominally the same peak. For example[6, 7], in KCl the divalent impurity linked F centres anneal at 85°C, the sodium impurity F_A centres dissociate at 120°C, F centres near dislocations anneal at 135°C and the 'normal' F centres anneal at 190°C.

(viii) The F centre is a halogen ion vacancy, so when it is formed under irradiation conditions we expect to see a complementary interstitial centre produced at the same rate. Such a centre exists, the H centre, and at low temperatures it is stable and has the expected properties (see § 7.3).

Excited states of the F centre

On the high-energy tail of the F absorption band one finds four very weak bands which have been labelled K, L_1, L_2, L_3 as shown in Fig. 7.5. Although weak they occur as a group and their strengths are each a constant fraction of the F band. For this reason Lüty[8] proposed that these represent transitions to higher states of the F centre. There are distinct experimental problems when working with such weak bands because of the ever present uncertainties of under-

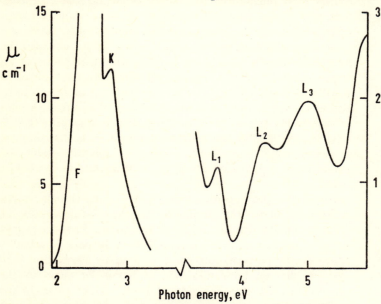

Fig. 7.5 The excited state absorption bands of the F centre in KCl.

lying bands and stray impurity bands. However, if these are excited states of the F centre, we might expect the following properties:

(i) The ratio of the absorption coefficients of K : L : F will be fixed under all conditions of preparation.

(ii) Changes in the F concentration, say by conversion to F′ centres, will be accompanied by changes in K and L bands.

(iii) They will follow a Mollwo-Ivey law.

(iv) They should give photoconductivity.

At first conflicting evidence was quoted, but there was no certainty that this did not merely result from other weak absorption processes which fall in the same spectral region. Finally a convincing experiment by Chiarotti and Grassano[9] showed that the bands are related to the same ground state as the F centre.

Their experiment was to modulate the normal F centre ground state population by an intense slowly fluctuating light source and simultaneously measure the changes produced in the rest of the spectrum with a second weak light beam. The first excited state population of the F centre is also modulated at the same frequency, but the maximum of the population is out of phase with the population of the ground state. The results show that the F, K, L_1, L_2, L_3 bands are all modulated in phase, so have a common ground state. The weak F′ band which extends across this entire region of the spectrum is in phase with the upper state of the F centre as expected for reactions involving electron transfer from excited F centres (i.e. F* + F→ F′ + α).

Recent band structure calculations[10] predict the correct positions for all the excited states with only a one-parameter fit. The calculation also predicts that the upper states lie in different conduction bands even though these bands overlap. Evidence which supports this has come from experiments on the movement of the peak positions under hydrostatic pressure[11]. The change in peak energy with pressure dE/dP depends on the curvature of the bands and may be positive or negative depending on the particular transition. The fact that the F, K, L_1, L_2, L_3 bands differ in both sign and magnitude of dE/dP suggests that the upper states lie in different bands.

The F′ centre

We have already mentioned that this consists of an F centre plus a second weakly bound electron. The small attractive potential for this electron means that the wave-functions extend over several atomic neighbours away from the centre. By comparison with the F centre this produces a broader, lower energy absorption band whose width is insensitive to the temperature changes, since many atoms are involved and the atomic energy levels are shallow. The centre is thermally and optically unstable and is in equilibrium with the F centres by a reaction

$$2F \rightleftharpoons F' + \text{vacancy}.$$

Optical bleaching from one to the other involves a passage via the excited state, and here we might write (F′)* + α→2F*→2F + 2$h\nu_F$ to describe energy losses by F centre emission. The efficiency of conversion between the two centres is a function of temperature and is shown in Fig. 7.6.

On bleaching the centre or exciting it to the upper state, the loosely bound electron is freed, so we observe strong photocurrents. The temperature response of the photoconductivity is not as simple as suggested in Chapter 5, because both the retrapping and the conversion efficiencies are temperature-dependent.

Finally, this is a centre with two trapped electrons, so it is unlikely to produce ESR signals for the ground state, and the excited state

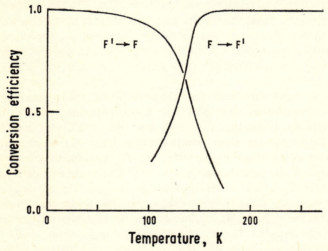

Fig. 7.6 The quantum efficiency for the interconversion of F to F^1 centres by optical bleaching in KCl.

studies are complicated by the weak binding, so that the electron is easily released.

M centres

For an M centre model such as that shown in Fig. 7.1 we may make some positive tests of the model.

(i) Because the centre is two adjacent F centres, we expect any conversion between F and M centres will proceed as $2F \rightleftharpoons M$. This is true for additive colouring experiments and is approximately true for conversion by optical bleaching. Here there are secondary problems because of competing reactions to form the other related F$'$, R and N centres.

(ii) The centre has a dipole axis along <110> and can be optically reoriented, so a polarized light irradiation will bleach some <110> centres and change the balance of the six <110> oscillator directions. This will produce both anisotropic absorption and luminescence. The analysis of these data was discussed in § 3.3.

(iii) The ground state has two electrons, so there is no ESR signal. However, in this case the first excited state is still localized, so studies may be made on the higher state by resonance methods[12]. Both ESR and ENDOR measurements show that the defect has a total spin of unity and the line positions are consistent with a pair of adjacent overlapping F centres.

(iv) A more generally important consequence of a centre such as this which has inversion symmetry is that it should not have an electric dipole moment. The fact that no electric dipole moment could be detected was a strong argument in favour of this model at a time when no ESR measurements had been made on the excited state.

(v) Finally one may also check that there are no unpaired spins in the ground state of the centre by magnetic susceptibility measurements. Determination of the total susceptibility of the sample as a function of M centre concentration has advantages over ESR measurements, because there are no problems of power saturation or overlapping resonances.

R centres

The model shown in Fig. 7.3 is not the only arrangement of three F centres which can exist, but polarized absorption and fluorescence measurements, together with ESR measurements of the triplet state, suggest that this model with a <111> dipole axis is most appropriate. Claims of up to six R absorption bands have been made, and this implies that not all the details of the three F arrangements are yet known.

7.3 Hole-trapping centres

There are many possible hole traps, but the two centres which are best understood are the V_K and the H centre. They are also prototypes of hole centres in other materials, so we will mention them again (see also Chapter 2). The V_K centre is a hole self-trapped in the lattice on a pair of halogen lattice ions. The H centre is a halogen atom which is in a crowdion interstitial position in a row of halogen ions.

The models shown in Fig. 7.7 have excellent verification from the magnetic resonance techniques and also occur in many other compounds which include halogen ions. This is because the central halogen molecular ion is extremely stable. Of the other hole traps, labelled as V centres, we know that these produce absorption bands such as those at 230, 270 and 410 nm in KBr and 240 and 355 nm in KCl, but the detailed structures are not settled although they are also related to interstitial halogen atoms. Because most of the hole traps produce absorption features in the same spectral region, one must be particularly precise in quoting the band positions or ensure that only one absorption band is present. Failure to do this has caused some confusion in the labelling of V_1, V_2, etc., to the same features.

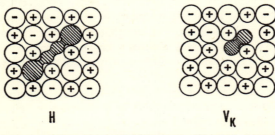

Fig. 7.7 The models for the V_K and H centres. The hole is trapped on the shaded ions in each case.

The V_K centre

It is surprising that this defect is stable, since it consists of two normal lattice ions which have trapped a hole. However, this is possible because the lifetime of the excited state is longer than the time required for the ions to relax into the molecular ion configuration. Consequently the unit is relatively stable, for example, up to 150 K in most alkali halides. Despite the immensely detailed knowledge that we have of this centre from ESR and ENDOR, it is instructive to consider what other properties this defect will have.

(i) The defect has <110> symmetry, and because it involves only normal lattice ions the hole can easily be transferred from one pair to another. This allows the centre to be reoriented with polarized light. This is possible even at 4 K.

(ii) Because no impurity ion movement is required, the centre will thermally disorient, i.e. cease to remain in a preferentially bleached set of directions. This will occur at a lower temperature than is required for annealing.

(iii) On annealing, the hole will be released so will give thermo-luminescence and thermally stimulated currents.

(iv) It is also apparent that, in order to make the defect, holes must be liberated in the lattice. Under normal conditions of radiation electrons and holes are produced in pairs, so the rate of V_K formation will be influenced by the number of electron traps since these inhibit electron-hole recombination. Thus, addition of electron traps such as Tl, Pb, or Ag in KCl will result in an enhancement of the rate of V_K growth with irradiation.

(v) Conversely the annealing stage will require free electrons to combine with holes liberated from the V_K centres. This is re-flected in the annealing temperature, and whereas in pure KCl the V_K annealing stage is around 130 K, in lead doped KCl the apparent annealing stage is delayed until 220 K. In fact this may measure the stability of the electron trapped at the impurity.

(vi) This defect is basically a halogen molecular ion perturbed by a crystal lattice. Such a molecule has two optical transitions and the bands in KCl occur at 365 nm in the ultra-violet and at 750 nm in the infra-red. The successful prediction of the position of more than one optical absorption band is a valuable step in the test of a defect model.

(vii) As a final check on the model we can look at the magnetic resonance data, since the defect has a single trapped hole. We find that it is equally coupled to the two halogen ions, and a determination of the g value, symmetry and nuclear hyperfine interactions shows this to be the correct model of the V_K centre[13].

The H centre

Referring to Fig. 7.7, we see this centre is a halogen neutral atom which has been inserted into a row of halogen negative ions to form a crowdion-type interstitial. From the changes produced in the row this appears as a central strongly coupled pair of halogens which form a halogen molecular ion on one lattice site and a weakly bonded halo-gen ion on either end of the pair. Clearly the central pair is dominant in the defects' properties, and it is not appropriate to consider the system as being composed of four equal ions. With this in mind we shall now discuss some of the expected properties of the centre.

(i) The central molecular ion is the same as the V_K centre, so we expect optical absorption bands at similar energies. However, the tighter bonding in the H centre may move the bands to higher energies. For example, in KCl the V_K aborption band was found at 365 nm (3·4 eV) and the H band occurs at 345 (3·6 eV).

(ii) The defect contains an extra atom, compared with the pure lattice, so it is likely to be an efficient hole trap, and thus the production rate will not be influenced by impurities in the lattice, in contrast to the V_K centre.

(iii) Similarly the energy of the annealing stage is insensitive to impurities.

(iv) During irradiation the interstitial is formed at the same rate as halogen vacancies, so both the F and H asborptions bands for new defects should grow at the same rate.

(v) During annealing not only will charge be released from the defect to give thermoluminescence or thermally stimulated currents, but also the extra ion must find a new site. We therefore expect a rise in related absorption bands as the H centre is destroyed.

(vi) The <110> symmetry and hole character are essentially the same as for the V_K centre, so we expect the same type of ESR signal. The weakly coupled pair of ions at the ends of the H centre chain will modify the spectrum by splitting each resonance line into further hyperfine terms from the additional nuclear spins. This, of course, was found[14], and an example of V_K and H spectra is shown in Fig. 7.8 (see also Chapter 2). In this result the strong F centre resonance could not be suppressed so the line exists in both curves.

7.4 Impurity centres

In many ways the identification of impurity centres is simplified because chemical analysis, or intentional crystal doping, indicate which ions are of particular importance. So the choice of a defect model merely becomes an exercise in deciding between substitutional, interstitial, and associations of impurity-intrinsic or impurity-impurity complexes. As usual, ESR techniques will be fruitful if the impurity ion has nuclear interactions to give hyperfine structure lines to the spectrum. An impurity ion can have other effects than just the production of a new absorption band. This often presents some problems in selecting which properties result directly from the impurity and which result from changes in the number of electron or hole traps. For instance, in the previous discussion of the V_K centres we mentioned that Pb doping of KCl altered the annealing tempera-

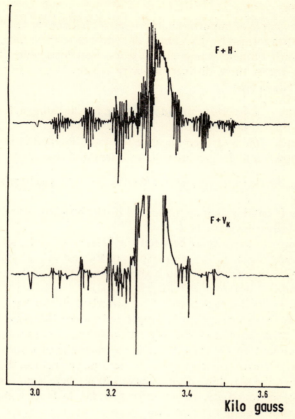

Fig. 7.8 A comparison of the ESR signals from KCl crystals containing F and V_K centres and F and H centres.

ture of the V_K absorption from 130 K to 220 K. There is obviously the potential error that one might interpret this result as evidence for a lead absorption band which happens to overlap the V_K absorption band but anneals with a higher activation energy. This mistake is unlikely when well established defects are involved, but in a new class of material such problems are not trivial. Similarly in the case of centres formed by a combination of an intrinsic and an extrinsic defect the changes in the peak position of the modified intrinsic centre may be interpreted as resulting from totally new defects. Such effects were discussed in the production of F_A and Z centres.

U centres

As an example of impurity centres in alkali halides we shall discuss the U centres formed by the additon of hydrogen to the lattice. This is a particularly suitable example because it is well documented and the hydrogen forms three distinct impurity centres. The resultant absorption bands are:

U centre —a hydride ion substituting for a halogen ion on a lattice site.

U_1 centre—the hydrogen negative ion in an interstitial site.

U_2 centre—a hydrogen atom in an interstitial site.

Among the list of properties and interactions of these bands are the following:

(i) The U band is directly influenced by the host lattice and follows a Mollwo-Ivey law of $E \propto d^{-1.10}$.

(ii) There are similarities between the U and F bands, since they occupy the same site with respect to the rest of the lattice. The similarity extends even to higher excited states such as the U_K band (i.e. notation of K, L_1, L_2, L_3).

(iii) The U band bleaches with light absorbed in the band. However, the centre can be destroyed in two possible ways. Firstly the hydride ion may lose its electron and secondly the hydrogen atom may escape. The difference between the two is that in the case of electron release a charge motion can occur in the lattice. When the bleaching occurs above 170 K, there is no associated photoconductivity, so we infer that the hydrogen atom was released during the bleaching stage.

(iv) The high temperature bleaching follows a reaction

$$U + h\nu_{(U)} \rightarrow U_2 + F$$

where the U_2 centre arises from a neutral hydrogen interstitial and the defect remaining at the lattice site is an F centre.

(v) At lower temperatures a different reaction is produced by the bleaching light and the entire hydride ion becomes interstitial and does not thermally dissociate. The final defects are the U_1 centre and an empty halogen vacancy, i.e.

$$U + h\nu_{(U)} \rightarrow U_1 + \alpha$$

(vi) Photoconductivity can be produced by exciting the U centre into a higher excited state via the U_K band.

(vii) Interconversion between the interstitial defects is also possible by the reaction

$$\alpha + U_1 + h\nu_{(U)_1} \rightarrow U_2 + F.$$

(viii) In these reactions the rate-controlling process is the rate of diffusion of the hydrogen away from the U centre. So the theory may be tested by substituting deuterium for the hydrogen and noting a corresponding reduction in the bleaching rate because the diffusion coefficient is less for the heavier isotope.

(ix) Magnetic resonance is again useful for the atomic interstitial hydrogen in the U_2 configuration but is unsuitable for either the U or U_1 cases where there are two electrons in the outer shell of the atom.

The U centre has been the subject of much theoretical work[15, 16, 17], because firstly it has only two electrons and secondly it resembles the F centre. Even a simple approach to the calculation of the energy levels using a mixture of a $1s$ hydrogen ground state and the $2p$ states of the F centre gives a reasonable estimate of the absorption band energy. More detailed studies of the vibrational spectra are interesting because of the large polarizability of the hydride ion. Consequently there are major changes in the wave-functions of the states during vibration. Infra-red measurements of the vibrational spectra have been made by Schaefer[18]. Additional information from these vibrations has been used to explain why there are three annealing stages for the U centre[19]. The interpretation is that in the temperature range 100 to 230 K one successively sees the annealing to close pairs, medium range and totally free interstitial hydrogen and vacancy sites.

7.5 Defect studies in other materials

In principle no further examples of defects are necessary, because the simple alkali halide prototypes have already set the pattern of study for other materials. While this is true, it may be more satisfying to realize that the problems can be solved in other systems and also that the time involved in the development of a defect model is now greatly reduced because of the number of precedents. Generally the presence of an ESR signal simplifies the problem, or at least provides a more conclusive result. Occasionally the number of resonances is embarrassingly large, such as the 200-odd line spectra one finds in gamma-ray-irradiated sodium chlorate, but even here the lattice symmetry can be used to simplify the interpretation of the spectra.

When considering the study of a new compound, it is desirable to use a material which is chemically and physically stable and which can be formed into pure single crystals with regular faces. However, when there has been a particular interest in an explosive, deliquescent or unstable compound these experimental problems have been faced. Consequently the literature includes defect studies even on such materials as lead azide, lithium iodide or silver bromide.

We cannot over-emphasize the need to start with pure material, preferably in crystalline form. Unintentional impurity studies have already consumed many hundreds of man-research years, and although it may be possible to identify the basic defects (such as F or M) other studies have little value. For example, in the alkali halides the rate of defect production during irradiation by an X-ray beam can vary by a factor of a hundred, depending on the supplier and the vintage of the crystal.

Pure crystal preparation is still a difficult art because impurity levels of consequence to defect studies should be measured in parts per billion, not in normal chemically detectable quantities. A good example of a systematic approach to the problem is given in Chapter 9, where we describe the work of the group at the Oak Ridge Laboratory to improve the quality of KCl crystals. Similarly the semiconductor industry has also made great progress in purification. However, their applications generally require that the material be 'electrically pure', and 'inactive' impurities may exist in large quantities. Such a limited definition of purity may prove disastrous in a study of defects and their interactions.

The following short list of compounds in which positive defect identification has been made is not intended to include all the data available, and we apologize to those authors whose work we have accidentally or intentionally ignored.

The alkali azides

The first class of materials that we shall consider are the alkali azides, because they are a classic example of materials which should follow the alkali halide prototype. Because of difficulties in preparing the samples they have received very little attention. The crystal structure is similar to the alkali halides, but as shown in Fig. 7.9 the N^-_3 azide radical replaces the halide ion and is set at an angle of 45° to the side of the unit cell in order to make the lattice more compact. This results in a tetragonal unit cell which for KN_3 has sides of 0·609 nm and 0·706 nm. The quality of the crystals obtainable is still very poor, because the materials thermally decompose at a temperature below their melting point. They also undergo photolysis (i.e. de-compose when illuminated). This means that growth techniques cannot involve high temperatures (except under very high pressures of a nitrogen atmosphere in a pressure bomb) and zone refining is not possible. Preparation from solution is still an unsatisfactory approach

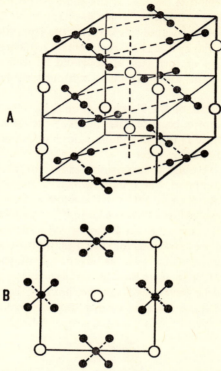

Fig. 7.9 The crystal structure of potassium azide. The potassium ions are on the $\frac{1}{4}$, $\frac{3}{4}$ planes and the nitrogens are in the 0 and $\frac{1}{2}$ planes. The figures show a perspective view A and a projection on the [001] plane B.

except for samples intended for the simplest of experiments. Finally one should probably conduct the growth in darkness. These problems are not insoluble, and if there were commercial demand for pure single crystals the problems could be solved.

Because of the quality of the crystals only a few basic defect studies have been done. Nearly all the successful defect identification has come from spin resonance which requires only small single crystal, or powder, samples[20, 21, 22, 23, 24]. Several variants of the F centre can occur, because the azide radical contains three ions. The simple F centre, an electron trapped in a N_3^- vacancy, has been seen by ESR and optical absorption in both NaN_3 and KN_3. The defect occurs after irradiation with X-rays or ultra-violet light, but it is not known whether the formation process is the same as in alkali halides. It would seem unlikely that the Pooley mechanism (to be described in Chapter 8) operates, because this involves the separation of the vacancy and interstitial by a $<110>$ replacement sequence and the azide ions are set skew to this axis, so it would be difficult to transfer momentum or atoms along the chain without large energy losses at each collision.

The ESR pattern of the F centre should be very similar to the alkali halide F centre, because it has precisely the same environment. The only difference is that in sodium azide one actually resolves all 19 lines of the hyperfine spectra.

Optically the F centre is surprising, because in NaN_3 it is *not* the largest absorption band. The F band, which correlates with the ESR centre, occurs at 730 nm, but is overshadowed by a pair of bands which can be resolved at 547 and 625 nm[25]. These have been interpreted as an F_2^+ centre (i.e. an F centre associated with an anion vacancy). There are two bands for such a centre because of the differences in symmetry possible with the linear azide radical. A list of the absorption bands with their possible assignments is given in Table 7.1. This should be treated as a very preliminary set of models until further work is done, because at this stage various authors[25, 26] differ on the choice of F or F_2^+ for the strongest band.

Table 7.1

	Wavelength, nm		
	F	or F_2^+	V
NaN_3	730	547, 625	330
KN_3	790	570	361
RbN_3	750 or 820	577	374
CsN_3	850	593	390

Other nitrogen centres seen by ESR are an interstitial nitrogen atom and a linear N_4^- ion at a lattice site and an N_2^- ion on an azide ion site. Finally results from thermoluminescence experiments[27] have been interpreted in terms of a V_K centre, and a study of infra-red vibrational spectra[28] suggests that a cyclic group of N_3 ions develops after irradiation.

The identification of colour centres in the azides has not been of primary interest, and most defect studies have concentrated on the rates of photolysis and thermal decomposition. However, these are related problems, because the sensitivity of a decomposing material can be controlled by suitable centres which act as catalysts. Suitable centres are often intrinsic defects such as colloidal metal centres. Control of these defects can offer control of the sensitivity of these thermally unstable, i.e. explosive compounds.

The alkaline earth fluorides

There are very close similarities between the alkali halides and the alkaline earth fluorides despite the difference in crystal structure produced by a divalent cation. The magnetic resonance measurements of both ESR and ENDOR have been very successfully applied, and the F, V_K and H centres have all been identified[29, 30, 31, 32]. Despite this there has been a surprising amount of controversy as to the identification of the optical absorption bands. The problem has arisen because the cations of Ca, Sr and Ba are very different in size (0·099, 0·112 and 0·134 nm), and this leads to a distortion of the space available for an F centre. From alkali halide results one intuitively expects the absorption bands to follow a Mollwo-Ivey law. However, the theoretical basis for this is that the potential well scales linearly with the lattice parameter, i.e. $V = V(r/d)$. A theoretical calculation using such a potential function[33], as in the point ion calculation, will of course predict a Mollwo-Ivey law. However, if one includes ion size effects[34] the Mollwo-Ivey law no longer applies. Thus according to one's choice of theoretical treatment the law predicts different positions for the F band. In this group of compounds the F bands in CaF_2 and SrF_2 were identified and one wishes to extrapolate to the F band in BaF_2. This is unwise as the 'law' is only suitable for interpolation. Further, there are sufficient colour centres produced by either doping or electrolytic coloration that one can always find data to fit the chosen law. The absorption band positions are as follows.

Table 7.2

	Wavelength, nm			
	Experiment	Theory		Lattice constant d
		Point ion	Ion size terms	
CaF$_2$	376	388	346	0·5451 nm
SrF$_2$	435	426	426	0·5784 nm
BaF$_2$	611 or 505	475	598	0·6187 nm

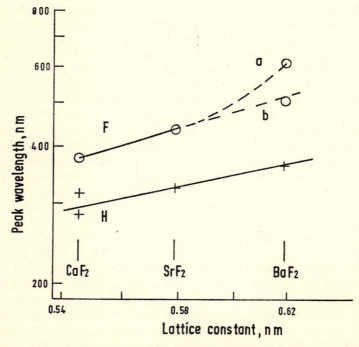

Fig. 7.10 A plot of peak wavelengths of possible F and H bands versus lattice parameter for alkaline earth fluorides. Curve (a) is the theoretical Mollwo-Ivey plot including ion size effects; (b) without ion size effects.

In this table we have included the predicted peak wavelengths of the F band for the simple point ion-type calculations and those which also include ion size perturbations. The data are plotted in Fig. 7.10 on a Mollwo-Ivey plot of log λ versus log d. The doubts are sufficiently great that from the above table one cannot choose between the 611 or 505 nm peak for the F band in BaF_2[35], but strong evidence for the 611 nm peak is provided by Hayes and Stoneham[36], who report only a 611 nm peak in samples which showed by ESR that they contained 10^{18} F centres cm^{-3}. Further support for the theory of Bartram et al.[34] is given by the changes in the hyperfine splitting constants in the ^2S ground state and the spin-orbit coupling in the ^2P excited state of the F centre.

The data for the H centres are also plotted on Fig. 7.10, but in this case the Mollwo-Ivey slope is 1·5. One expects the law to be valid for exponents within the range 1 to 2. A discussion of the present understanding of defects in alkaline earth fluorides has been given by Hayes[37]. The aggregate centres are more complex than in the alkali halide lattice, and for example the equivalent M band has at least two components. The peak positions quoted for both these, and the V_K and H centres, are listed in Table 7.3.

Table 7.3

	M bands		V_K	H
CaF_2	366	521	320	314 or 285
SrF_2	427	595	326	325
BaF_2	550	725	336	364

The other alkaline earth fluorides have different crystal structures (i.e. not cubic). In the tetragonal MgF_2 the peak at 260 nm has been identified with the F centre[38], and because of the crystal symmetry the band is slightly anisotropic. Also there are several possible configurations for an M centre formed from two F centres and the bands at 370 and 400 nm seem appropriate.

MgF_2 is interesting in that it also forms new defects under ionizing radiation as do the alkali halides, but the other alkaline earth fluorides do not do so if 'pure' or at room temperature.

The alkaline earth oxides

This series of compounds has many features which make them attractive for defect studies. They have a wide band gap, are chemically stable and crystallographically simple with the same structure as the alkali halides. Consequently a considerable amount of work has been done with these materials, both experimentally and theoretically, and most of this work is summarized in the review paper by Henderson and Wertz[39]. A disadvantage of such a strongly bonded system is that the compounds have very high melting points and pure crystal preparation is not very advanced. For example, Henderson and Wertz state that they have never found MgO which is free of iron impurities. In fact the iron group elements enter this series very easily as substitutional ions and exist in several valence states in the crystals. In many instances these impurities show ESR signals. The ESR parameters of Ag and the coupling constants for some 42 impurities in the MgO, CaO, SrO, BaO series are quoted in the review[39].

Since the ions are doubly charged, the anion vacancy centre can be in either the singly or doubly charged state, so we find both a 'one-trapped-electron' and a 'two-trapped-electron' absorption band. (N. B. In the later notation with respect to the crystal these are F^+ and F[39].) Also, as with the alkaline earth fluorides, there is controversy over the assignment of the F band in BaO because of the large cation. The possible bands are listed in Table 7.4 below.

Table 7.4

	Peak position, nm		
	Two e^-(F)	One e^-(F$^+$)	Lattice constant d
MgO	247	251	0·4203 nm
CaO	400	345	0·4797 nm
SrO	497	400	0·5144 nm
BaO	620	620 or 485	0·5523 nm

The two bands in MgO are very close together, but suitable bleaching techniques have been used to resolve them[40]. On a Mollwo-Ivey plot Turner[35] assumed that the point ion description is appropriate and chose exponents 3·2 and 2·4 for the two- and one-electron centres. However, the results of Bessent et al.[41] disagree with this. The question is not yet settled, and the two alternatives seem to be that: (a) as in the fluorides the Mollwo-Ivey law is not appropriate and the BaO F^+ peak is at 620 nm and possibly overlaps the F peak. Alternatively (b), the difference in lattice structure lessens the importance of the large cation in the oxides.

Hole, impurity and aggregate centres have also been detected in these compounds. Electron spin resonance measurements on other group II–VI materials have revealed that F centres exist in the entire series formed from Mg, Ca, Sr, Ba and O, S, Se.

Further examples of ESR data for the oxide structures of rutile, perovskite, spinel and garnet are reviewed by Low and Offenbacher[42].

Diamond

So far we have considered only colour centres in ionically bonded lattices. However, the alternative type of bond is exemplified by the covalently bonded diamond lattice and a model of the structure is shown in Fig. 7.11. As was mentioned in § 3.1, diamond is a wide

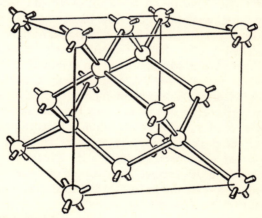

Fig. 7.11 The diamond lattice.

band gap insulator which has an indirect band gap of 7·02 eV and a direct gap of 5·47 eV. There are some problems in obtaining suitable single crystals for the defect studies, because natural samples tend to be rather small, i.e. millimetre dimensions, and the conditions of growth preclude satisfactory artificial production of large, pure, perfect crystals. Consequently all defect studies are made on the impure, natural samples. Depending on the quality of the transmission in the ultra-violet the diamonds are classified as type I or type II. The type I material is strongly absorbing whereas the type IIa samples are transparent down to 5·4 eV. Similar transparent samples which show semiconducting properties are termed type IIb.

For our purposes these classifications are unnecessary, since they only indicate the presence of different concentrations of impurity in the diamond. All these features and many of the defects which have been detected in diamond result from the presence of nitrogen impurities, and since this is a pentavalent ion it will act as a donor whether it occupies a substitutional or an interstitial site. The experimental evidence for the substitutional site is provided by spin resonance data[43, 44], by optical absorption, and by photoconductivity and electrical resistance measurements[45]. The photoconductivity only occurs for photon energies above 1·7 eV. The interpretation that the donor level is 1·7 eV below the conduction band is supported by the data on the changes in electrical resistivity with temperature. An Arrhenius plot (log σ versus $1/T$) also gives an activation energy of 1·7 eV. Theoretical studies of the orbitals for the donor electron of this centre[46] have correctly predicted the observed quadrupole and hyperfine splitting constants of the ESR lines.

Nitrogen in an interstitial site is apparent in the form of an absorption system labelled ND1 which is centred at 3·4 eV. This absorption spectrum is typical of absorption bands in diamond in that it has a sharp transition line, at 3·149 eV, which is only a photon-induced transition and is termed a zero phonon line. A series of overlapping lines on the high-energy side of the zero phonon line results from interactions with the possible combinations of lattice phonons. The resolution of the peaks is improved at lower temperatures and is further improved when one can remove the effects of overlapping absorption bands in this region. The example of some original data and the resolved spectra shown in Fig. 7.12 has been taken from the measurements of Davies and Lightowlers[47].

These very narrow absorption bands occur because in the covalently bonded system one finds highly localized defects. Such line spectra are ideal for determining the symmetry of the defect, since it is possible to predict the movement of the lines and the resulting polarization which will occur when the sample is stressed along various axes. It is also possible to use the method of moments[48], i.e. the shape of the absorption band, to note the changes in spin-orbit coupling as the defect site is distorted by a uni-axial pressure.

For the ND1 band the stress results agree with an interstitial model[47] for the defect, and the association with nitrogen is possible, because the strength of the band correlates with the presence of nitrogen.

Nitrogen also exists in other states in the diamond, and concentrations of up to 0·2 atomic per cent have been detected[49]. Some of the nitrogen exists as platelets in the (100) planes. These islands of nitrogen were initially estimated to be as large as 100 nm in diameter[50],

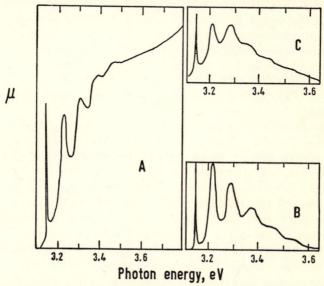

Fig. 7.12 Absorption bands produced in diamond by 2 MeV electron irradiation. Original data measured at 84 K (A). After subtraction of overlapping bands, measurements made at 84 K (B) and 273 K (C).

but more recent X-ray diffraction measurements[51] suggest that much of the nitrogen exists in small clusters, possibly only as pairs.

Intrinsic defect studies in diamond have been centred on the absorption band with a zero phonon line at 1·673 eV and higher order structure up to 2·8 eV. This series of bands was labelled the GR1 bands[52] (i.e. the green to red spectral region). The bands represent the lowest energy transition and the phonon-assisted transitions of the neutral carbon vacancy. They can be produced by fast electron radiation damage, and the calculated displacement energy (see Chapter 8, § 8.6) of 80 eV is consistent with the formation energy of an intrinsic lattice defect in a strongly bonded material. Unlike the alkali halides, where the vacancy centre (the F band) always appears after irradiation, the GR1 band, which is the equivalent vacancy centre in diamond, does not always appear. Although it is easily formed in the type I and IIa diamonds, it only appears in the p-type IIb diamonds after prolonged irradiation. If we realize that both type I and IIa diamonds contain nitrogen to such concentrations that they are n-type material, we can now explain this problem. The results imply that the GR1 centre is an intrinsic defect with a ground state so high in the energy gap that it is above the Fermi level in p-type material, and therefore without trapped electrons. In the type IIb diamonds prolonged irradiation[53] compensates the hole traps, and the Fermi level rises so that the GR1 band, and the ultra-violet nitrogen bands, appear. The earlier work of Clark et al.[52] on the changes in strength of the various absorption bands during annealing cycles is consistent with this interpretation.

Unfortunately the theoretical calculations for the symmetry and transition energies of the carbon vacancy centre are so complex that there is only limited agreement with the predicted and observed positions of the absorption bands[54-57]. For example, uni-axial stress experiments on the zero phonon line do not agree with the simple model for the defect, unless one assumes a suitable Jahn-Teller distortion of the defect site. These problems are discussed in the recent review paper by Clark and Mitchell[58] on radiation-induced defects in diamond, but because of the difficulties in the theoretical estimates of the position of the absorption band one can but state at this stage that the model of a carbon vacancy for the GR1 centre is most probable.

In conclusion it appears that defect studies in diamond are likely to be limited to work on impure specimens. The more general observation is that in the tightly bonded systems we may produce the characteristic sharp lines of the zero phonon absorption.

Sodalites

For our next example of defect studies we have chosen the class of materials known as the sodalites. There are several reasons for considering these materials even though the number of investigations that have been made up to now is relatively small. This material is the first that we have considered that has a complex unit cell with a large number of atoms. It is also one of the few substances that colours under ultra-violet or electron irradiation, and it is this latter property that has determined the interest in the material, because it is already used to a limited extent in information storage devices such as the dark trace oscilloscope (i.e. where the electron beam writes a dark line on a light background.)

Sodalites have a formidable structure, because the basic unit is comprised of sodium chloride in a matrix of sodium aluminium silicate, an ensemble which is chemically described by $6(NaAlSiO_4)$ $2NaCl$. Consequently a diagram such as Fig. 7.13 which is only

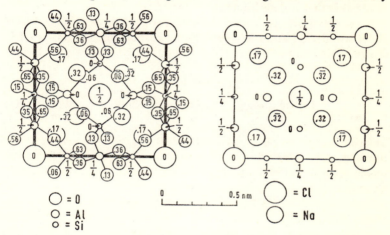

Fig. 7.13 The structure of sodalite. The cubic structure is projected along an axis (after R. W. G. Wyckoff, *Crystal Structures* (Interscience) Vol. 4, p. 430, 2nd edition, 1968).

two-dimensional does not give an ideal picture of the atomic structure, and to visualize the overall crystal and defect structure requires much more thought than for the alkali halides. It may be helpful to consider the structure as a framework of aluminosilicates which are joined by the oxygen ions, and the open cages which result contain four sodium ions surrounding a central chlorine ion. Crystal preparation is difficult, and one typically uses a hydrothermal process for the growth. Large single crystals have been prepared by Bye and White[59]. The material also occurs as a mineral in the form of hackmanite. One disadvantage of a hydrothermal method of growth is that water and other impurities are easily included in the final crystal, and of course with such a complex structure it is difficult to avoid the problems of non-stoichiometry or interchange of ions into other sites.

The motivation for the study of sodalites comes from the fact that the material can be made photochromic, i.e. that not only can it be coloured by the absorption of light, but also the process can be reversed by a further optical process. The application of photochromic materials to information storage or display systems will be described in Chapter 9. Although several materials show these properties the efficiency of bleaching and the writing speed are particularly favourable in the sodalites. The disadvantage of the currently available material is that it deteriorates with usage.

A strong visible absorption band is produced by electron or ultraviolet irradiation and the colour centre responsible for this is an F centre. This band at 530 nm also has an ESR signal with a 13-line

Table 7.5

Ion	Lattice spacing, nm	ESR hf line spacing, gauss	F band position, nm
Chloride	0·888	31·0	530
Bromide	0·9936	28·3	555
Iodide	0·9011	24·7	600
Ge in Si sites	0·9037	31·3	—
Ga in Al sites	0·8947	32·8	550
Hydroxy-sodalites	0·903	No F band formed	

hyperfine stectrum[60]. The defect is readily identified as an electron trapped in a chlorine ion vacancy, because only this site has four equal sodium neighbours. These are all Na^{23} atoms of nuclear spin 3/2, so we can have a maximum spin of six, which gives the 13-line hyperfine spectrum $(2I + 1)$.

The sodalite lattice is readily modified to accept other ions, and we can replace chlorine by Br^-, I^-, or OH^-, the aluminium by gallium and the silicon by germanium. There is a change in lattice spacing and the ESR and optical absorption bands also undergo small changes in energy[61]. Control of the peak energy is desirable in device manufacture, and here the F band can be moved from pink to blue by changing from chlorine to iodine ions. Some examples of the changes which result in the lattice when we alter the constituents are given in Table 7.5.

Although F centres are readily formed by ionizing radiation, the multiple vacancy centres have not been detected and are unlikely to exist in sodalites because of the very large unit cell. In the case of M centres in alkali halides the two adjacent F centres of KCl are only 0·22 nm apart, whereas in sodalite one would require a strong inter-action at a separation of 0·89 nm. The intervening structure makes this most unlikely.

The crystals do not necessarily exhibit photochromism after growth, and so various treatments to induce further defects to alter the Fermi level, or to rearrange existing defects have to be made before the material is photochromic. Successful empirical treatments include 'sensitizing' in a hydrogen or inert gas atmosphere for 30 minutes at 900°C. Alternatively one can intentionally dope the material with S, Se or Te, but because of the nature of the doping technique one is also adding O^-, O^{--} and OH^- unintentionally during the hydrothermal process.

The various treatments result in an ultra-violet absorption band near 4·7 eV, which can absorb light to generate the F band and in turn reap-pears following bleaching of the F band by F band light. The similarity of the bands at 4·7 eV following the various treatments suggests that the defect responsible for the band is an impurity linked to an intrinsic defect or possibly to an 'unavoidable' impurity in the crystal. Initially it was thought the defect was an S^{--} on a chlorine site which then donated the excess electron to an empty F centre during irradiation, but later claims[62] favour an F centre linked to an O^{--} ion.

Other radicals which have been detected in sodalite include O^-, O^{--} and SO_4^+. The site assigned to the SO_4^- ion is a chlorine vacancy and the O^- site occurs at the common apex of a pair of oxygen tetrahedra. Detailed measurements of the oxygen site suggest that the adjacent metal ions were both aluminium rather than the alternate aluminium and silicon which should occur in the perfect lattice.

This introduction to the defects in sodalite illustrates the way in which empirical and quantitative defect studies interact, and assist one another for a material which is under rapid development for a commercial application. For example, the defects produced by the sensitizing heat treatment are unknown, but the rationale of such a treatment to alter the Fermi level, charge state and vacancy concentration is obvious.

Glass

All the examples that we have considered so far have been in crystalline materials, but there is no *a priori* reason why we cannot observe colour centres in non-crystalline substances. Even by restricting the discussion to inorganic glasses, we find a vast range of compounds whose electrical and optical properties can be adjusted by the inclusion of defects or impurities. Although glass structures do not have long range order, the atomic arrangements do have short range order, which extends over several interatomic spacings. This is sufficient to produce a well defined environment for a defect in the glass, since we will recall that even in perfect single crystals we rarely needed to consider atomic interactions beyond next nearest neighbours. For example, in the case of the F centre the band position was only slightly altered in the F_A or Z arrangement. Of course, the properties determined by the crystal symmetry will not occur, and many of the interactions which aided specific identification of defect models, such as the nuclear interaction terms in ESR, will be of less value in the glass matrix. Similarly the statistical nature of the atomic spacing in the matrix will produce a small range of environments, so that optical absorption bands are likely to be somewhat broader and less sensitive to temperature in glasses compared with crystalline materials. All the preceding discussions of the effect of the Fermi level and impurities on trap population, and the interactions of traps by association or charge exchange, will be as appropriate in the amorphous material as in crystalline solids. Finally we should realize that many glasses are only metastable, and the presence of defects can aid the process of crystallization, or vitrification, within a glass.

Just as an example of the continuity between the crystalline and amorphous state we can cite the case of the V_K centre in alkali halides and alkali halide-borate glasses[63]. In both cases we find that one of the defects which is stable at low temperature is the Cl_2^- molecular ion. This is easily identified by the ESR spectrum and also gives two optical absorption bands. In the glass the binding between the chlorine ions and the trapped hole is somewhat stronger than in the crystal, because there are only weaker bonds to the neighbouring ions. One can sense this both by the interaction terms, which give Δg, and the hyperfine splitting of the ESR spectrum, and by the position of the optical absorption band, which will move to higher energies in the case of a more strongly coupled molecule. For example, the V_K absorption band occurs at 3·40 eV (365 nm) in KCl, but is moved up to 3·67 eV (338 nm) in a glass composed of 7·5% KCl, 22·5% K_2O, 70% B_2O_3.

There is a wealth of data accumulated on defects formed in glass matrices, and for further details and references one may consult the review papers by Lell *et al.*[64] and Bishay[65]. Some general guidelines that result from these studies should be mentioned. For example, in most of the borate, phosphate or silicate glasses the visible absorption bands result from hole traps. Any movement of the Fermi level which is produced by doping with trivalent ions, such as cerium, or by irradiation damage, can alter the intensities of these bands and even make some electron traps observable. In oxide glasses one also finds that irradiation damage destroys oxygen bonds which normally bridge two cations, and frequently new bonds form to isolated alkali ions. Similarly the introduction of higher valence impurity ions can locally modify the glass network and produce colour centres.

The range of constituents in a multicomponent glass is continuously variable, but one finds that the major absorption features occur in approximately the same region of the spectrum for all members of a series. For example, Bishay quotes the major bands of various glasses as:

Glass	Major absorption bands, eV		
Alkali borates	2·5	3·6	4·9
Cabal, CaO, B_2O_3, Al_2O_3	2·3	3·5	5·0
Soda lime silicate	2·9	4·0	5·5
Phosphate	2·3	2·9	5·5

Such similarities are to be expected, because the network of oxygen ions is producing similar interatomic spacings and environments, and also the bands are always broad because of the short range order.

We mentioned earlier that glasses are metastable and tend to form crystalline phases over a limited region of the matrix, and when this occurs there can be a major change in the properties of the region. Impurities and defects play a major role in initiating these changes. Despite the fact that the subject is not yet fully understood, we shall take as an example the changes that occur in chalcogenide glasses when they undergo a change of state and switch from a condition of electrical insulator to electrical conductor. These glasses are being intensively studied, because if one plots the current-voltage characteristic of the glass there is a condition at which the glass switches its conductivity state in times of the order of nanoseconds, a speed unmatched by any other device. There is a similar reverse switch and/ or a memory effect when the voltage is reduced. Obviously the device applications for such a switch are enormous.

A currently held view of the effect[66] is that thermal runaway occurs and filaments of high-temperature and high-conductivity material are formed. These same filaments may alter into an ordered glass structure. While it is certain that impurities and defects play a role in this transition, there is insufficient evidence at this stage to speculate on the actual mechanism. The magnitude of the problem is apparent, when we realize that the 'pure' chalcogenide glass may have a chemical composition approaching $As_{30}Te_{48}Si_{12}Ge_{10}$. One can safely predict that a considerable amount of work has yet to be done on this material.

Silicon

As a final example we will mention the identification of defects in semiconducting materials. Here the problems and experimental approaches are identical with those in the insulating materials, but there is frequently much more information on the electrical properties and hence the position of the defect energy levels within the band gap. As before, the conductivity, photoconductivity and optical absorption measurements may be used to separate properties of the various defects, but it is generally impossible to identify the specific centres without magnetic resonance data. The semiconductors also show the property that intrinsic defects can be produced by fast electron bombardment (see § 8.6) and in the case of a compound semi-conductor one might hope to separate the production of vacancies in the two sub-lattices by irradiation with different energy electrons.

Silicon has been particularly well studied, and a list of the major defects and the appropriate ESR parameters are recorded in the book by Corbett[67]. The silicon lattice readily accepts tri- and pentavalent ions substitutionally to form donor or acceptor sites in the material, so it is reasonable that we should find silicon vacancies in either plus or minus charge states. These are not always immediately apparent if the Fermi level is too low in the band gap, as it may be from stray impurities, but this situation can be changed by excitation with light In such cases there is then evidence for a vacancy with two electrons in it.

The normal tetrahedral bond arrangements for these diamond-type lattices are modified to form bridging molecular bonds for the atoms around a vacancy[68]. This is apparent for the case of the divacancy in silicon which is shown in Fig. 7.14. In this figure the absence of an adjacent pair of ions has produced the formation of two 'bent' bonds for the two closest pairs of silicons, and interaction across the defect region for the other pair of silicons. This defect was analysed more readily, because one could align the defect by an applied stress and hence reduce the number of possible models for this centre. As with the simple vacancy, we expect the divacancy to exist in several charge states, and both the one- and three-electron spectra have been seen by ESR[69, 70].

Higher order complexes of vacancies into groups of three or four have been proposed to explain other resonance data[67]. Also the

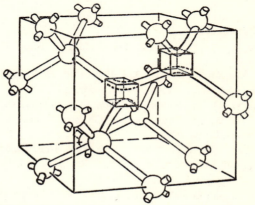

Fig. 7.14 A model of a divacancy in silicon. The vacancies are shown by cubes and the bent bond directions are those which result from the defect.

association of impurities with vacancies is established for complexes such as a silicon vacancy adjacent to substitutional oxygen, phosphorus or aluminium. We also expect that during a radiation damage event interstitials will be formed. These are highly mobile so are stable only at low temperatures, but with irradiations made at 4 K they were successfully identified[68] in impure silicon. Surprisingly the impurity appears as the interstitial rather than a silicon ion, which suggests that the original silicon interstitial is able to substitute back into the lattice at an impurity site.

7.6 Conclusion

Colour centre studies are possible in all insulating materials no matter how complex is their structure or unstable is their chemistry. It is encouraging to realize that all the materials which have been studied have revealed details of some of the defects, and there is no reason to suppose that this advance cannot be continued. The one major step in all these studies which must be made, and is the most frequently ignored, is to ensure that the starting samples are made to the highest possible specifications. Failure to do this has all too often resulted in pointless experiments.

The theoretical studies of defects also involve major problems, and the discrepancies between alternative calculations indicate that considerable care should be used when applying the theoretical predictions.

Chapter 7 References

[1] Markham, J. J., 'F Centers in Alkali Halides', Supp. **8**, *Solid State Physics*, 1966.

[2] Seitz, F., *Rev. Mod. Phys.* **18**, 384, 1946; ibid. **26**, 7, 1954.

[3] Sonder, E. and Sibley, W. A., to be published.

[4] Compton, W. D. and Rabin, H., *Solid State Physics*, **16**, 121, 1964.

[5] Lüty, F., *Physics of Color Centers*, Chapter 3, edited by W. Beall Fowler (Academic Press) 1968.

[6] Arsenovici, L. C. and Townsend, P. D., *Phil. Mag.* **25**, 381, 1972.

[7] Jain, S. C. and Mehendru, P. C., *Phys. Rev.* **140**, 957, 1965.

[8] Lüty, F., *Z. Phys.* **160**, 1, 1960.

[9] Chiarotti, G. and Grassano, U. M., *Phys. Rev. Letters* **16**, 124, 1966.

[10] Dawber, P. G. and Parker, I. M., *J. Phys. C.* **3**, 2186, 1970.

[11] Brothers, A. D. and Lynch, D. W., *Phys. Rev.* **164**, 1124, 1967.

[12] Seidel, H. and Wolf, H. C., *Physics of Color Centers*, Chapter 8, edited by W. Beall Fowler (Academic Press) 1968.

[13] Castner, T. G. and Kanzig, W., *J. Phys. Chem. Solids* **3**, 178, 1957.

[14] Kanzig, W. and Woodruff, T. O., *J. Phys. Chem. Solids* **9**, 70, 1958.

[15] Spector, H. N., Mitra, S. S. and Schmeising, H. N., *J. Chem. Phys.* **46**, 2676, 1967.

[16] Wood, R. F. and Öpik, U., *Phys. Rev.* **162**, 736, 1967.

[17] Wood, R. F. and Gilbert, R., *Phys. Rev.* **162**, 746, 1967.

[18] Schaefer, G., *J. Phys. Chem. Solids* **12**, 233, 1960.

[19] Fritz, B., *J. Phys. Chem. Solids* **23**, 375, 1962.

[20] Carlson, F. F., *J. Chem. Phys.* **39**, 1206, 1963.

[21] Carlson, F. F., King, G. J. and Miller, B. S., *J. Chem. Phys.* **33**, 1266, 1960.

[22] Gellerinter, E. and Silsbee, R. H., *J. Chem. Phys.* **45**, 1703, 1966.

[23] Horst, R. B., *J. Phys. Chem. Solids* **23**, 158, 1962.

[24] Wylie, D. W., Shuskus, A. J., Young, C. G., Gilliam, O. R. and Levy, P. W., *Phys. Rev.* **125**, 451, 1962; *J. Chem. Phys.* **34**, 1499, 1961.

[25] King, G. J., Miller, B. S., Carlson, F. F. and McMillan, R. C., *J. Chem. Phys.* **35**, 1442, 1961.

[26] Papazian, H. A., *J. Phys. Chem. Solids* **21**, 81, 1961.

[27] Townsend, P. D., *J. Phys. C.* **2**, 1116, 1969.

[28] Bryant, J. I., *Spectro Chim. Acta* **22**, 1475, 1966.

[29] Hayes, W. and Stott, J. P., *Proc. Roy. Soc.* **A301**, 313, 1967.

[30] Twidell, J. W. and Hayes, W., *Proc. Phys. Soc.* **79A**, 1295, 1962.

[31] Marzke, R. F. and Mieher, R. L., *Phys. Rev.* **182**, 453, 1969.

[32] Tantan, G. A., Shatas, R. A.,Williams, J. E. and Mukerji, A., *J. Chem. Phys.* **49**, 5532, 1968.

[33] Bennett, H. S. and Lidiard, A. B., *Phys. Lett.* **18**, 253, 1965.

[34] Bartram, R. H., Stoneham, A. M. and Gash, P., *Phys. Rev.* **176**, 1014, 1968.

[35] Turner, T. J., *Sol. Stat. Comm.* **7**, 635, 1969.

[36] Hayes, W. and Stoneham, A. M., *Phys. Letters* **29A**, 519, 1969.

[37] Hayes, W., *Rad. Effects* **4**, 239, 1970.

[38] Facey, O. E. and Sibley, W. A., *Phys. Rev.* **186**, 926, 1969.

[39] Henderson, B. and Wertz, J. E., *Adv. Phys.* **17**, 749, 1968.

[40] Kappers, L. A., Kroes, R. L. and Hensley, E. B., *Phys. Rev.* **B1**, 4151, 1971.

[41] Bessent, R. G., Cavennett, B. C. and Hunter, I. C., *Phys. Chem. Solids* **29**, 1523, 1968.

[42] Low, W. and Offenbacher, E. L., *Solid State Physics* **17**, 135, 1965.

[43] Smith, W. V., Sorokin, P. P., Gelles, I. L. and Lasher, G. J., *Phys. Rev.* **115**, 1546, 1959.

[44] Loubser, J. H. N. and DuPreez, L., *Brit. J. App. Phys.* **16**, 457, 1965.

[45] Farrer, R. G., *Sol. Stat. Comm.* **7**, 685, 1969.

[46] Every, A. G. and Schonland, D. S., *Sol. Stat. Comm.* **3**, 205, 1965.

[47] Davies, G. and Lightowlers, E. C., *J. Phys. C.* **3**, 638, 1970.

[48] Henry, C. H. and Slichter, C. P., *Physics of Color Centers*, Chapter 6, edited by W. Beall Fowler (Academic Press) 1968.

[49] Kaiser, W. and Bond, W. L., *Phys. Rev.*, **115**, 857, 1959.

[50] Evans, T. and Phaal, C., *Proc. Roy. Soc.* **A270**, 538, 1962.

[51] Sobolev, E. V., Lisoivan, V. I. and Lenskaya, S. V., *Sov. Phys. Doklady* **12**, 665, 1967.

[52] Clark, C. D., Ditchburn, R. W. and Dyer, H. B., *Proc. Roy. Soc.* **A234**, 363, 1956; ibid. **237**, 75, 1956.

[53] Dyer, H. B. and Ferdinando, P., *Brit. J. Appl. Phys.* **17**, 419, 1966.

[54] Coulson C. A. and Kearsley, M. J., *Proc. Roy. Soc.* **A241**, 433, 1957

[55] Yamaguchi, T., *J. Phys. Soc. Japan*, **17**, 1359, 1962.

[56] Lanoo, M. and Stoneham, A. M., *J. Phys. Chem. Solids*, **29**, 1987, 1968.

[57] Ritter, J. T., *Sol. Stat. Comm.* **8**, 773, 1970.

[58] Clark, C. D. and Mitchell, E. W. J., *Rad. Effects*, **9**, 219, 1971.

[59] Bye, K. L. and White, E. A. D., *J. Cryst, Growth*, **6**, 355, 1970.

[60] Hodgson, W. G., Brinen, J. S. and Williams, E. F., *J. Chem. Phys.* **47**, 3719, 1967.

[61] McLaughlan, S. D. and Marshall, D. J., *Phys. Letters* **32A,** 343, 1970.

[62] Ballentyne, D. W. G. and Bye, K. L., *J. Phys. D.* **3**, 1438, 1970.

[63] Griscom, D. L., *J. Chem. Phys.* **51**, 5186, 1969.

[64] Lell, E., Kreidl, N. J., and Hensler, J. R., *Prog. Ceram. Sci.* **4,** 1966.

[65] Bishay, A., *J. Non-Cryst. Solids* **3**, 54, 1970.

[66] Male, J, and Warren, A., *New Sci.* **47,** 128, 1970.

[67] Corbett, J. W., 'Electron Radiation Damage in Semiconductors and Metals', Supp. **7,** *Solid State Physics*, 1966.

[68] Watkins, G. D., *Radiation Damage in Semiconductors* (Academic Press) p. 97, 1965.

[69] Watkins, G. D., Corbett, J. W. and Walker, R. M., *J. Appl. Phys.* **30,** 1198, 1959.

[70] Watkins, G. D. and Corbett, J. W., *Phys. Rev.* **138**, A543, 1965.

8

MECHANISMS OF DEFECT
FORMATION

8.1 A survey of the mechanisms of defect formation

In any study of colour centres we will wish to know how to control the number and type of defects formed in the solid. Therefore, not only must we identify the defects, but we must also understand the mechanism of defect formation. This is a more challenging problem than the identification, because it is a dynamic event with a time scale for radiation damage mechanisms in the range 10^{-6} to 10^{-12} seconds.

Listed below are several alternative routes that lead to defect-rich material. Of course, not all of them are suitable for the production of all types of defect, and many treatments are further limited to certain classes of materials.

The common techniques are:

(i) Quenching of materials from a high temperature where there is a thermodynamic equilibrium concentration of defects. Rapid cooling can freeze in this equilibrium.

(ii) Control of the crystal growth to include impurity atoms or produce changes in the stoichiometry.

(iii) Chemical additions to an existing crystal by additions from the surface. Typical methods are diffusion of a surface layer, ion exchange, fast ion implantation, or nuclear reactions within the solid.

(iv) Ionic motion induced by internal breakdown of the crystal at high electric fields.

(v) Physical displacement of atoms from lattice sites by collisions with high-energy particles (i.e. momentum transfer).

(vi) Low-energy mechanisms which transfer energy to a lattice (i.e. little momentum in the incident particle or photon).

For simplicity we shall discuss these topics in terms of specific crystals. However, the methods, and mechanisms involved, are equally applicable to non-crystalline materials such as glasses or polymers as is evident from the literature. It will also become apparent that many of the above methods are applicable to metals as well as insulators.

8.2 Quenching

Simple intrinsic defects such as vacancies, divacancies or interstitials may reasonably be expected to occur in all materials if the energy of formation U_F is not too high. Such defects may remain in the lattice, if the sample is cooled to a temperature at which annealing cannot occur. To estimate the concentration of defects (n/N_0), where N_0 is the number of lattice sites, we may assume that statistical thermodynamics is appropriate, so that

$$\frac{n}{N_0} = \exp\left(-\frac{U_F}{kT}\right).$$

We can estimate the formation energy of a vacancy from a very simple model by considering the following sequence of events. First we remove an atom from the bulk of the crystal by breaking the chemical bonds, and then we replace it on the surface. If the 'bulk' atom has twice as many bonds as the 'surface' atom, then the net expenditure of energy is the same as that for removing a surface atom. In other words it is the energy of sublimation. So the formation energy of a vacancy is around 1 to 2 eV.

At a high temperature, say 1160 K, the defect concentration for an E_F of 1·0 eV becomes

$$\frac{n}{N_0} = \exp\left(\frac{-1\cdot0}{8\cdot6 \times 10^{-5} \times 1160}\right)$$

$$\approx 4\cdot5 \times 10^{-5}.$$

This is an appreciable fraction of the lattice sites, with the defects being, on average, some forty atomic spacings apart. If we now cool the specimen slowly so that it remains in thermodynamic equilibrium, no vacancies will remain, but a rapid quench will freeze in the vacancies. The fraction of defects retained depends on the speed of the quench, the rate of defect coalescence and the number of sinks for the vacancies. Hence it depends on the impurity and dislocation concentrations. For metals, high cooling rates of 10^4 K per second are feasible, but for insulators (i.e. low thermal conductivity) this is impossible because the strain introduced under the thermal shock may shatter the specimens. Despite this, the method will still retain sufficient defects to affect optical absorption or electrical conductivity.

8.3 Impurities and non-stoichiometry

A convenient treatment of impurities and defects which is used to predict their concentrations is to consider them as chemical species. We thus write transitions of one type of defect into another as chemical reactions and predict the final concentration from the law of mass action.

In this instance free electrons or holes are also chemical entities which must be included. The analysis is particularly useful in reactions of a solid with a vapour phase.

A simple bulk 'chemical' reaction may be the conversion of α centres to F centres by electron capture,

$$\alpha + e^- \rightleftharpoons F.$$

The appropriate mass action law of concentration is

$$\frac{[\alpha][e^-]}{[F]} = K, \text{ where } K \text{ is a constant.}$$

Now consider a more complex example in which a compound AB is in contact with gas A. In the solid, A can exist as normal lattice atoms, neutral interstitials, charged molecules or vacancies. Possible reactions are

$$A_{solid} \rightleftharpoons A_{gas} + A_{vacancy}, \quad \frac{[A_{gas}][A_{vacancy}]}{[A_{solid}]} = K_1,$$

$$A_{solid} + e^- \rightleftharpoons A_{interstitial}, \quad \frac{[A_{solid}][e^-]}{[A_{interstitial}]} = K_2,$$

$$A_{interstitial} + A_{interstitial} + e^- \rightleftharpoons A^-_{molecular\,ion}, \quad \frac{[A_{interstitial}]^2[e^-]}{[A^-_{molecular\,ion}]} = K_3.$$

The external gas pressure determines $[A_{gas}]$ and the number of lattice sites is large and constant so the equations become

$$P_A[A_{vacancy}] = K_1',$$

$$\frac{[e^-]}{[A_{interstitial}]} = K_2',$$

$$\frac{[A_{interstitial}]^2[e^-]}{[A^-_{molecular\,ion}]} = K_3.$$

Solution of these simultaneous equations gives the defect concentrations. From a set of equations such as this we expect the gas pressure to control the number, type and charge state of the defects.

In compounds such as rutile (TiO_2) which can readily be made non-stoichiometric by reduction in a vacuum furnace, we can control the degree of non-stoichiometry by the partial pressure of oxygen on the sample during reduction. In rutile there are two possible ways of building a crystal which is deficient in oxygen. The first is to maintain the titanium atoms at the lattice sites and have some oxygen lattice sites unoccupied. The second way is to construct a complete TiO_2 lattice and disperse the additional titanium as interstitials.

The kinetics of the possible reactions has been well studied in this material, and one predicts that a measurement of titanium interstitials (by electrical conductivity σ) depends on oxygen gas pressure

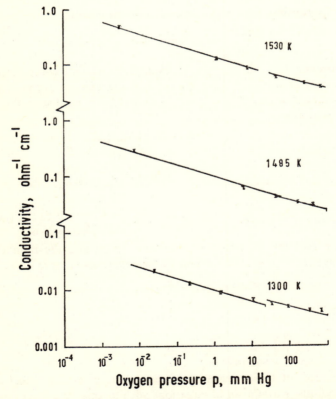

Fig. 8.1 A plot of the electrical conductivity of rutile at 1300, 1485 and 1530 K as a function of the surrounding oxygen pressure. The lines represent slopes of $-\frac{1}{5}$ or $-\frac{1}{6}$.

as $P^{-1/5}$. The alternative oxygen vacancy model would give $\sigma \propto P^{-1/6}$. Initially there were conflicting results, but Yahia[1] found that both pressure dependences exist for a limited range of temperature. Results shown in Fig. 8.1 for the conductivity measurements at 1300 K give $\sigma \propto P^{-1/6}$ for pressures above 10 mm of Hg pressure (i.e. oxygen vacancies), but $\sigma \propto P^{-1/5}$ at lower pressures (i.e. titanium interstitials).

In semiconductor manufacture, control of this balance between defects also gives control of the type of conductivity. A compound semiconductor such as lead sulphide is strongly electron-type if exposed to a low vapour pressure of sulphur, and hole-type after treatment at high sulphur pressures.

This type of conductivity is frozen in when the sample is cooled to room temperature, and so forms a clean method of altering the intrinsic conductivity.

8.4 Chemical doping of crystals

From the preceding sections it is apparent that chemical impurities can control the electrical and optical properties of a crystal. They also modify the equilibrium concentration of intrinsic defects and the rate at which new defects are formed by irradiation. The former is demonstrated by the addition of impurities of different valence to a material (i.e. As or P into Ge). Vacancies can also be stabilized in this way; for example, the addition of divalent calcium to a compound like sodium chloride produces vacancies, for the calcium occupies a lattice site and the vacancy can compensate the additional charge on the divalent ion.

The effects of impurities on the rate of radiation-induced absorption bands was discussed in § 3.7. Common examples are the addition of sulphur or dyes to photographic emulsion. Possibly less well known is the need to stabilize many high explosives, such as lead azide, which dissociate (explosively) by either thermal or irradiation treatment. In this example the incoming radiation produces electron-hole pairs, which causes rupture of the chemical bonds and the release of chemical energy. If the rate of release is sufficiently great the process accelerates and the sample explodes. Presumably the introduction of impurities provides trapping sites for one type of charge, and although defects are formed the rate is sufficiently reduced that the chemical energy is dissipated as heat to the surroundings before the sample is destroyed.

It is impractical to dope from the vapour phase if the compound decomposes or undergoes structural changes, so many chemicals are added to an existing material at relatively low temperatures.

Material deposited on the surface will migrate into the bulk at a rate dependent on the ion size, temperature, potential barriers and mechanism of diffusion. If atoms interchange with lattice and interstitial sites the total activation energy will be the sum of the formation energy E_F, and the migration energy E_M. This will give a diffusion coefficient D as a function of temperature,

$$D = D_0 \exp - \left(\frac{E_F + E_M}{kT} \right).$$

With material originating on the surface of the sample the flux of ions moving into the bulk depends on the concentration gradient. This is expressed by Fick's law as

$$f = - D \frac{dN}{dx}.$$

Thus the depth distribution will depend on the temperature and time (i.e. the number of random jumps). In real materials extended defects such as dislocations and grain boundaries present a low-energy path into a sample for incoming atoms. In the early stages of diffusion the extra atoms decorate the dislocations and the effect is commonly used to render the dislocations visible.

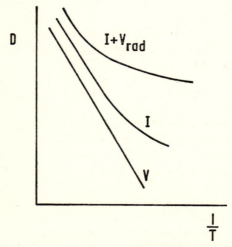

Fig. 8.2 Changes in the vacancy diffusion coefficient with temperature for vacancies only V additional impurities I, and $(I + V_{rad})$ impurities plus radiation.

Diffusion which takes place by an interchange of impurity and vacancy sites is not a random process, and the lattice strain produces a preferred direction for the jump of the impurity towards the vacancy. Such correlated jumps lower the diffusion coefficient.

This also implies that the diffusion coefficient for both the impurity and the vacancy will change in a sample which has been irradiated to produce additional vacancies. In Fig. 8.2 the diffusion coefficient of an intrinsic vacancy is seen to be a simple exponential function of temperature in a pure material. Impurities raise the diffusion rate at low temperatures, and radiation-enhanced diffusion occurs when additional vacancies are introduced.

Qualitatively we expect rapid diffusion for small ions entering a lattice, and the rate will be anisotropic if the crystal structure contains open directions. For example, lithium entering a quartz lattice can easily move down the open channels of the c-axis.

There is a similarity between random diffusion of ions into a lattice which exists when the driving force is a concentration gradient and the diffusion which occurs when ions are electrically driven into a sample. The two cases are related by the Einstein expression

$$D = \frac{kT\mu}{q}$$

where q is the charge on the ion and μ is the ion velocity under unit electric field (i.e. the mobility). The electrical conductivity is proportional to the charge concentration and mobility ($\sigma = Nq\mu$), so it is related to the diffusion coefficient by

$$\sigma = \frac{Nq^2 D}{kT}.$$

Consequently impurities, or impurity-vacancy complexes, are important in determining the electrical properties of insulators. For wide band gap insulators all the electrical conduction can be from ionic motion, as is shown by Fig. 8.3 for KCl doped with $BaCl_2$.

Ion exchange can also take place at the surface between the material and a surrounding gas or liquid. Experimentally this may be simpler than depositing a surface layer of one element, but the diffusion process into the bulk is the same for both systems.

8.5 Ion implantation in a solid

There is an upper limit to the amount of material that can be introduced into the host material before the structure undergoes a

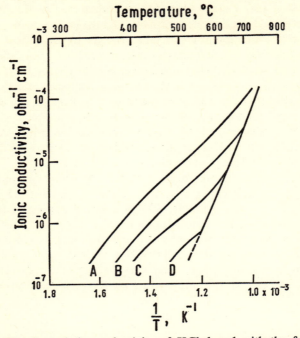

Fig. 8.3 The electrolytic conductivity of KCl doped with the following concentrations of BaCl$_2$.
A = 14 × 10^{-5}; B = 4·67 × 10^{-5}; C = 1·9 × 10^{-5}; D = 'pure'.

phase change or the impurities form colloidal clusters. This solubility limit may be so low that it is impossible to dope with particular elements. Solubility considerations, but not clustering, can be overcome by the technique of ion implantation. Ion implantation is a powerful technique which has the following advantages.

(i) The implantation is not limited to any particular type of ion.

(ii) The crystal may be kept at any chosen temperature during or after the implant.

(iii) The depth distribution is controllable by a suitable choice of the ion energy.

(iv) The experiment may be made under high-vacuum (i.e. clean) conditions and, if necessary, only one isotope is used.

(v) Several different isotopes may be implanted at the same, or different, parts of the sample.

(vi) Very accurate spatial control of the implantation is ideal for device manufacture. This is made possible because there is only a small lateral spread in the ion beam in the solid. Thus the process lends itself to complete automation by controlling the position and energy of the ion beam.

(vii) Finally the inherent radiation damage produced during implantation can be annealed at relatively low temperatures so that the spatial control of the implant is not lost.

The subject has been excellently reviewed for ion beam doping of semiconductors by Dearnaley[2] and optical applications have also been suggested elsewhere[3, 4].

With the advent of nuclear reactors theoretical treatments to evaluate the defect concentration which results from the passage of energetic particles through solids have become necessary and have been developed. Much of this theory is relevant to ion beam implantation in a solid and is described in standard texts on radiation damage[5, 6, 7].

For our purposes we need only summarize the factors to be considered without justifying the choice of the repulsive potential between ions or the approximations in finding the total defect concentration.

For a head-on collision an ion of mass M_1 and energy E_P can transfer an energy E to an atom of mass M_2, given by

$$E_{\max} = \frac{4M_1 M_2}{(M_1 + M_2)^2} E_P.$$

This will displace the struck atom from the solid if $E_{\max}$ is greater than the binding energy of the atom to the lattice site. The appropriate energy E_D is not the sum of the bond energies but an energy some four or five times greater, because the displacement is a dynamic event which occurs in a time short compared with the lattice relaxation time. Hence more energy is required to force the struck atom from its site. (An energy of 25 eV is frequently used as a first guess for most solids.) Both the first particle and the struck atom may recoil with energies far in excess of E_D, so an entire region becomes disrupted. One estimate of the number of displacements produced by the nuclear collisions is (see, for example, reference 5)

$$n = \tfrac{1}{2}\left[1 + \ln\!\left(\frac{E_{\max}}{E_D}\right)\right].$$

An alternative is $\qquad n \approx 0.35\dfrac{E_p}{E_D}.$

The ions travelling through a solid lose energy either by electronic excitation or by direct collisions with other nuclei. Each mechanism of interaction involves both the ion sizes, charge state and velocity. If we ignore the subtleties of the theories we can describe the relative contribution to the loss by each mechanism by Fig. 8.4. For an ion of atomic mass 25 entering a solid of average atomic mass 25 the peaks in the energy loss curve will occur near 20 keV for nuclear collisions and 20 MeV for electronic interactions.

In the high-energy range the energy loss rate is proportional to the ion's velocity, despite the changing state of ionization as the particle

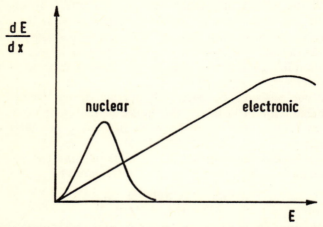

Fig. 8.4 A comparison of the nuclear and electronic mechanisms of energy loss for ions of different energy in a solid.

is slowed down. Lindhard and Scharff allow for these changes and estimate that the upper peak occurs near $(Z_1^{\frac{2}{3}}e^2)/\hbar$.

The low-energy curve from nuclear (i.e. hard sphere) collisions is sensitive to the choice of repulsive potential and the problem is discussed by Lindhard et al.[8].

From Fig. 8.4 we see that a fast ion entering a solid initially loses energy to the lattice and does not suffer nuclear collisions until it is approaching the end of its path. Consequently the depth at which the implant takes place is a function of the energy. Similarly the straggling in ranges about this value is also energy and ion mass dependent. The most probable range is predictable from the Lindhard theory, e.g. Fig. 8.5, and the range distribution is shown for krypton ions fired into amorphous aluminium oxide in Fig. 8.6.

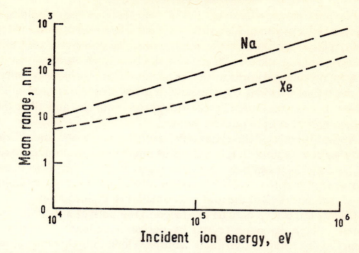

Fig. 8.5 A log-log plot of the calculated mean range of sodium and xenon ions in aluminium oxide.

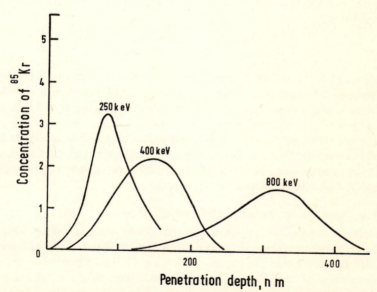

Fig. 8.6 Range of krypton ions of various energies implanted into alumina. (Data of Jespergard, P. and Davies, J. A., *Can. J. Phys.* **45,** 2983, 1967.)

For amorphous solids we should note that ion implantation, at one energy, can produce a fairly well controlled penetration into the solid. This is true even if the solid is unstable and has a protective coating, because the ions do not initially lose much energy.

A totally new phenomenon exists in regular lattices if the symmetry of the lattice provides directions with a low density of atoms. The incoming ions may be aligned with these channels, and then they penetrate very deeply into the lattice.

As an example of the enhancement of the range one finds[9] that on a 40 keV krypton implantation of amorphous aluminium oxide 1% of the ions travel beyond 35 nm, whereas, for a similar ion beam directed along the $<110>$ crystal directions of the crystalline Al_2O_3, at least 1% of the ions penetrate more than 250 nm.

Finally impurities may be produced in the bulk of a carefully prepared, chemically pure, crystal by nuclear reactions. This may be achieved by using an unstable isotope which subsequently decays or, more normally, by an induced radioactive decay during an irradiation. For bulk changes neutron activation is convenient, since the range and nuclear capture cross sections can be very large. Two such examples which both produce helium in the solid are the reactions

$$^{10}B + n \longrightarrow \alpha + {}^{7}Li,$$
$$^{25}Mg + n \longrightarrow \alpha + {}^{22}Ne,$$

although both produce helium ions (i.e. alphas) and an impurity isotope, the cross section for the neutron capture by boron is 3990 barns whereas the magnesium capture cross section is only 0·27 barn.

Alternative reactions are possible, such as $^{28}Si + \gamma \rightarrow {}^{27}Al + p$, but the technique is limited by the number of possible reactions in the conveniently accessible energy range and the fact that several decay products may be produced. In addition the sample may become too radioactive for normal experimentation.

8.6 Point defect formation by fast electrons

There is considerable interest in the simple intrinsic defects which exist when atoms are displaced from lattice sites to interstitial or surface sites within the lattice. Without a detailed knowledge of the properties of the simple vacancies and interstitials it would be illogical to attempt to identify complex defects.

In order to study the vacancy and interstitials in isolation from more complex defects, we choose to prepare 'pure' defect-free specimens and then selectively knock atoms from the lattice sites to interstitial sites. We must do this within the framework that there are no impurity or stoichiometry changes, that only one type of atom is displaced in a compound material, that the energy is only sufficient to move one atom and not cause divacancies or a disordered region, and finally that no diffusion is allowed (i.e. a low-temperature sample).

All these requirements can be fulfilled when the irradiation damage is made by relativistic electrons.

At energies in the range 0·5 MeV to 2·0 MeV the electron has sufficient momentum directly to displace lattice atoms but insufficient momentum for multiple damage events. Even with a compound system there will exist an energy region in which the energy transferred during a collision is above the threshold for displacement of the lighter atom but is insufficient to displace the heavier.

Such energetic electrons have a relativistic mass, and their energy is

$$E_1 = (m - m_0)c^2$$

where $m = m_0\left(1 - \dfrac{v^2}{c^2}\right)^{-\frac{1}{2}}$, v being the electron velocity. In a direct collision the maximum energy E_m transferred to an atom of mass M_2 is

$$E_m = \frac{2E_1(E_1 + 2m_0c^2)}{M_2c^2}.$$

Most collisions are not direct, so to compute the total probability of displacing an atom we use the differential cross section $d\sigma$ for Rutherford scattering of fast electrons through an angle θ given by

$$d\sigma\,(\theta) = \left(\frac{Ze^2}{2mv^2}\right)^2 \operatorname{cosec}^4 \frac{\theta}{2}\, d\Omega.$$

The solid angle of this annular cone is

$$d\Omega = 2\pi \sin \theta\, d\theta$$
$$= 4\pi \sin \frac{\theta}{2} \cos \frac{\theta}{2}\, d\theta.$$

We also realize that these are relativistic electrons, so

$$d\sigma\,(\theta) = \left(\frac{Ze^2}{m_0c^2}\right)^2 \frac{\pi}{\beta^4\gamma^2} \cos \frac{\theta}{2} \operatorname{cosec}^3 \frac{\theta}{2}\, d\theta$$

where $\beta = \dfrac{v}{c}$, $\gamma = \dfrac{1}{\sqrt{1 - \beta^2}}$.

Unfortunately we are treating the electron as a classical particle, which is only true for a de Broglie wavelength short compared with the distance of closest approach of the electron to the atom. To satisfy this condition we require $\dfrac{Ze^2}{\hbar c} > 1$. Evaluating this term gives $\dfrac{Z}{137}$ $(= \alpha)$, which is much less than unity. A more formal treatment for the electron scattering has been made by Mott[10], and the quantum mechanical cross section is related to the classical Rutherford scattering by $d\sigma_{QM} = R_s d\sigma_{classical}$. The factor R_s has an approximate analytical form given by McKinley and Feshbach[11]

$$R_s = 1 - \beta^2 \sin^2 \frac{\theta}{2} + \pi\alpha\beta \sin \frac{\theta}{2}\left(1 - \sin \frac{\theta}{2}\right).$$

Hence the scattering cross section becomes

$$d\sigma\,(\theta) = \frac{\pi Z^2 e^4}{m_0^2 v^4}(1 - \beta^2)\left[1 - \beta^2 \sin^2 \frac{\theta}{2} \right.$$
$$\left. + \pi\alpha\beta \sin \frac{\theta}{2}\left(1 - \sin \frac{\theta}{2}\right)\right] \cos \frac{\theta}{2} \operatorname{cosec}^3 \frac{\theta}{2}\, d\theta.$$

To relate this angular dependence to energy, we note that the energy transferred at an angle θ is less than that for a head-on collision by

$$E = E_m \sin^2 \frac{\theta}{2}.$$

Thus the differential cross section for energy transfer becomes

$$d\sigma\,(E) = \frac{\pi b^2 (1 - \beta^2)}{4} \frac{E_m}{E^2}\left[1 - \beta^2 \frac{E}{E_m} + \pi\alpha\beta \left\{\left(\frac{E}{E_m}\right)^{\frac{1}{2}} - \frac{E}{E_m}\right\}\right] dE$$

where $b = \dfrac{2Ze^2}{m_0 v^2}$. If the threshold energy for displacements is E_D, we obtain the total cross section for displacement by integration from E_D to E_m, i.e.

$$\sigma = \frac{\pi b^2 (1 - \beta^2)}{4}\left[\frac{E_m}{E_D} - 1 - \beta^2 \ln\left(\frac{E_m}{E_D}\right) \right.$$
$$\left. + \pi\alpha\beta\left\{2\left(\left[\frac{E_m}{E_D}\right]^{\frac{1}{2}} - 1\right) - \ln\left(\frac{E_m}{E_D}\right)\right\}\right].$$

If the electron energy is close to the threshold energy for displacement, then only the primary event produces damage, and the above formula approximates to

$$\sigma = \frac{\pi Z^2 e^2 (1 - \beta^2)}{\beta^4 \, ^2 m_0{}^2 c^4} \left(\frac{E_m}{E_D} - 1 \right).$$

Typical cross section curves are shown in Fig. 8.7 for a variety of

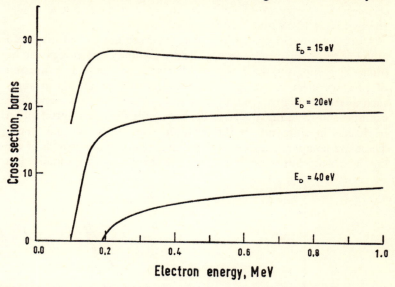

Fig. 8.7 Examples of the calculated cross sections for atomic displacements in carbon for fast electron bombardment. Curves are shown for different assumed displacement energies.

displacement energies of carbon atoms as a function of the primary electron energy. The subject is developed extensively in the book by Corbett[12] with the effects of secondary defect formation included in the analysis. To compute the total defect concentration and its depth distribution we require the range energy relation for relativistic electrons. Various empirical expressions have been used; for example, Glendennin[13] gives a range R for unit density material in units of mg/cm^2 for MeV electrons as

$$R = 407 E^{1.38}, \qquad 0.15 < E < 0.8 \text{ MeV},$$
$$R = 542 E - 133, \qquad E > 0.8 \text{ MeV}.$$

Alternative relations by Katz and Penfold[14] and Kobetich and Katz[15] take the forms

$$R = 412 E^{(1.265 - 0.0954 \ln E)}, \qquad E < 2.5 \text{ MeV},$$
$$R = 530 E - 106, \qquad E > 2.5 \text{ MeV},$$

and $R = 537E\left(1 - \dfrac{0 \cdot 9815}{1 + 3 \cdot 123E}\right),$ $\qquad$ $0 \cdot 3 \text{ keV} < E < 20 \text{ MeV}.$

As an example of the range a 2 MeV electron is stopped in 2·38 mm in Al_2O_3, which has a density of 3·97 g cm^{-3}. There is some uncertainty on the range of very low energy electrons, but a typical energy loss is 10 eV nm^{-1} in unit density material.

A crucial number for the calculation of the number of displaced atoms is the displacement energy. Experimentally this energy can be found by comparing the computed displacement curve as a function of primary electron energy with an observed parameter change, such as resistivity or the strength of an optical absorption band. The example shown in Fig. 8.8 gives the curves for the rate of production of defects in diamond for threshold energies of 20, 60 and 80 eV. These are compared with experimental results for the growth of the optical absorption band at 2·0 eV. The curves, when normalized at

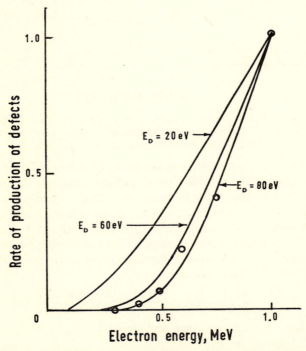

Fig. 8.8 A comparison of the normalized displacement cross section curves with experimental results for diamond. The optical absorption band at 2·0 eV was used as a measure of the damage.

an electron energy of 1·0 MeV, show that the high value of displacement energy is appropriate in this material. Similar results have been obtained with the conductivity changes produced in semiconducting diamonds. It is important to note that the curve shape must be used rather than the threshold energy at which damage becomes detectable. The latter approach may merely reflect the sensitivity of the measuring technique, particularly if minor processes exist which can cause some sub-threshold events.

One minor process which can occur in compound materials is the transfer of energy from the electron to a heavy atom via collisions with a light atom. The closer the masses (m, M_1, M_2) the more efficient is the transfer process.

As an example consider Al_2O_3, which has ions of mass 27 and 16. If the displacement energy for aluminium is 40 eV then a direct collision requires a primary electron energy of 0·385 MeV. However, the more efficient two-stage collision in which an electron strikes an oxygen ion which in turn strikes an aluminium ion will, in the optimum case, deliver 40 eV to the aluminium starting from a 0·29 MeV electron. Work on colour centre formation in sapphire (Al_2O_3) has indeed shown that there is a finite rate of production of the 6·1 eV

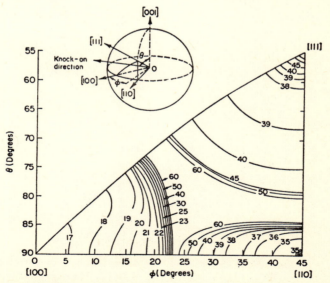

Fig. 8.9 Contours of constant displacement threshold in the triangle bounded by <100>, <110> and <111> directions in iron.

optical absorption band even at 0·3 MeV, although the threshold energy predicted from curves like those of Fig. 8·8 is above 0·38 MeV.

8.7 Displacement energies in real lattices

A lattice is not isotropic, so the energy required to displace an atom from the lattice site will depend on the direction in which the atom moves. The effect will be most pronounced in an anisotropic lattice, which has open channels as well as close-packed directions.

Detailed calculations of the displacement energy as a function of direction are shown rather concisely in the work of Erginsoy et al.[16] for iron. The contours of equal displacement energy about three crystalline axes are plotted in Fig. 8.9. The calculation also emphasized the role of replacement collision sequences as a means of separating vacancies and interstitials.

Because electron irradiation transfers most energy in a direct forward collision, the number of defects formed in thin samples directly reflects this anisotropy of E_D. In thin samples the electron

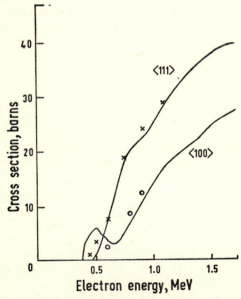

Fig. 8.10 Theoretical displacement cross section curves calculated for $E_D(100) = 20\,\text{eV}$ and $E_D(111) = 30\,\text{eV}$. The experimental results in iron for <100> irradiation are shown as ○ and for <111> as ×. The theory and measured curves were normalized at 1·1 MeV.

beam does not diverge appreciably and is essentially monoenergetic, so one can detect different damage threshold energies for different directions. Results of this type of measurement in iron[17] are shown in Fig. 8.10.

8.8 Low-energy mechanisms of defect formation

Whereas it is clear that irradiation of a solid with particles of high energy and large momentum can displace lattice atoms, it is not self-evident how defects can be formed by low-energy photon irradiation. Many materials undergo photolysis (i.e. dissociate during photon irradiation), but the detailed mechanisms of the energy transfer are far less well understood than the permanent defects which remain.

Loss of surface atoms during irradiation may occur by different processes than those which occur in the bulk, but it is these bulk mechanisms which are of interest here. Since direct collision mechanisms are inappropriate, the damage must result from ionization or excitation of lattice atoms. We shall see that it is currently thought that the key step for damage formation in alkali halides and photographic emulsions is the formation of an exciton. The energy associated with this bound electron-hole pair is then used to allow atomic movements. The alternative theories involving single or multiple ionization of the lattice ions have not yet been substantiated for the alkali halides. However, they should not be discarded, because they may still be applicable to other materials and they have certainly contributed to the recent excitonic models. For this reason we shall present the mechanisms suggested for defect formation in the alkali halides in a chronological order. The comments on the success or shortcomings of the models will be brief, and a more complete discussion of the background literature is to be found in the review article by Crawford[18]. Subsequent to the models for alkali halides we will present two models for the formation of defects in silver bromide.

8.9 The Seitz model

For historical reasons the Seitz[19] model was only attempting to show how X-rays could form vacancies in alkali halides. But it is noteworthy that Seitz began by considering the formation of an exciton. This was free to move through the crystal until it reached a dislocation line. Here the energy was to be deposited at a jog on a dislocation line with the resultant climb of the jog and the evaporation of a pair of vacancies into the bulk of the crystal.

Such a mechanism would produce a high defect concentration in the vicinity of the line, but at low temperatures there would be little diffusion into the bulk. If the mechanism operates then one expects the lattice to expand by vacancy injection and noticeable movement of the dislocation lines. Further, the process might proceed faster in crystals with a high dislocation concentration.

8.10 The Varley mechanism

Varley[20] approached the problem differently and proposed that the first step was multiple ionization of a halogen ion. Consequently the ion which had previously carried a net negative charge now became positively charged and was ejected from the lattice site by the electrostatic repulsion of its neighbours. With this mechanism one expects bulk defect formation of both halogen vacancies and interstitials and a threshold energy for multiple ionization which would be around 20 eV (i.e. very low energy X-rays). This model does not describe the subsequent fate of the interstitial or suggest how far the components of the pair are separated at low temperatures, which is a failing for alkali halide lattices, but need not be a problem in a more open lattice. One should also note that the vacancy and interstitial are not initially formed in the conventional charge state of an F centre and H centre but require charge capture to reach these configurations. In the process of doing this they may pass through higher excited states, so one expects luminescence corresponding to the transitions of the F centre.

For the process to be significant one also requires a high probability for multiple ionization and a relatively long-lived excited state for the multiply charged halogen ion. These considerations produced criticism and variations of the original Varley mechanism.

8.11 The Klick model

Dexter[21] pointed out that the multiple-ionized halogen was unstable against hole capture from one of its positive neighbours and should make this electronic change in about 10^{-15} seconds. This time is far too brief for the multiple-charged ion to be ejected from its site. Klick overcame this problem by a sequence of events shown in Fig. 8.11. In this model we change from a perfect lattice, Fig. 8.11(a), to a lattice with a multiple-charged halogen, Fig. 8.11(b); the following step is an instant capture of a hole from a neighbouring atom to produce a pair of neutral halogens, Fig. 8.11(c). Parenthetically we see that we could also have reached this condition by simultaneous single ionization of a pair of adjacent halogen ions.

There are a range of possibilities open at this point, and the stability of each was calculated by Williams[22]. He concluded that atomic motion is feasible before there is any further separation of the holes. Klick[23] had suggested a pairing of the neutral atoms to

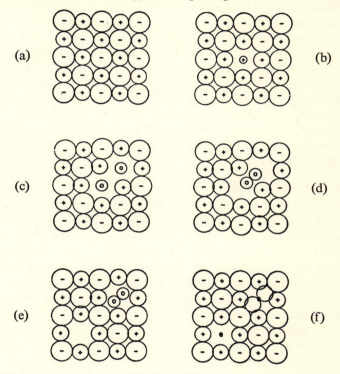

(a) (b) (c) (d) (e) (f)

Fig. 8.11 Schematic view of the Klick method of ionization of an alkali halide which leads to defect formation, as described in the text.

form a molecule, Fig. 8.11(d), and then a diffusion of this molecule away from the defect site by hole migration along the line of halogens. However, Williams considered the halogen molecular ion (say Cl_2^-) would probably be formed in preference to the neutral molecule. He then suggested the chemical energy released in this process could initiate a collision sequence along the line of halogen ions, that is along the close packed $<110>$ direction, Fig. 8.11(e).

Either scheme will result in a final stage of the type shown in

Fig. 8.11(f), but for stability the interstitial-vacancy pair should be further separated.

With this mechanism we expect defect formation at all temperatures, equal concentrations of vacancies and interstitials, and separations of the pair by a few atoms along a $<110>$ direction. The threshold energy for the process is that for multiple ionization or single ionization, if two events occur simultaneously side by side.

As an alternative to the multiple ionization Williams also calculated the force between a single halogen atom and a neighbouring halogen ion. The possible electronic levels allow a molecular ion to form, and possibly the molecule gains some energy from the Jahn–Teller relaxation of the neighbouring atoms from the defect. Hence it might be possible to start the replacement collision sequence with a single ionization event. The quantity of kinetic energy released by the chemical conversion is in some doubt, but the starting energy for the process must still be sufficient for ionization (i.e. at least the energy of the band gap).

8.12 The Pooley model

Measurements of luminescence and defect formation rates as a function of temperature led both Pooley[24] and Hersh[25] to propose basically the same model for the formation of the alkali halide F and H centres during irradiation with either X-rays or low-energy photons. It was observed that there was an anti-correlation between the rates, which suggests that the incident radiation is passed to the lattice in either of two ways, one of which gives luminescence and the other produces displacements.

It is proposed that the essential entity in this energy conversion is a molecular halide ion which undergoes excitation to form a bound electron-hole pair. One can draw a configurational co-ordinate diagram for this system of the form shown in Fig. 8.12, where the abscissa is a function of the separation of two halogen ions. The ground state equilibrium corresponds to atoms in their normal lattice sites. To estimate the energy required to reach the upper state, we need only argue that since the electron is still bound to the centre the energy is less than that of the band gap. In fact the work of Murray and Keller[26] suggests that the excited state is a normal exciton level.

In this figure the energy levels of the electron-hole pair are plotted as a function of the separation of two normal lattice ions which have self-trapped the hole part of the pair by an inward relaxation. This means that the initial separation of the charges occurs from a normally spaced pair of ions by exciton absorption, but the return to the ground state will depend on the thermal energy available to move the pair of ions together. At low temperatures a small relaxation enables the recovery to proceed by a radiative transition, but at high temperatures the ground and excited states may become mixed and the return path for the electron will be non-radiative and so the

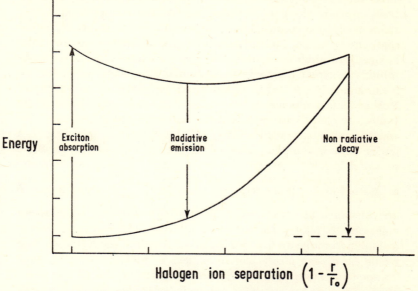

Fig. 8.12 The configurational co-ordinate diagram of the ground and excited states for an electron-hole pair in Kl as a function of the separation of the two halogen ions which have trapped the hole part of the electron-hole pair.

decay scheme will transfer kinetic energy to the pair of ions. This is the energy used to produce atomic displacements.

At this stage in the model we should note that: (i) the threshold energy is less than the valence-conduction band gap. (ii) The long-lived excited state, before decay, allows momentum conservation by interaction with the lattice, so the negligible momentum of the photon is unimportant in the movement of ions. (iii) Luminescence or defect formation are competing processes. (iv) The model has a calculable temperature dependence for luminescence and defect formation which is different for each alkali halide. (v) When the kinetic energy

CCS—N

is released the electron has already returned to the ground state. Therefore one expects colour centres to be formed complete with an electron in the ground state, rather than by a later stage electron capture or decay from excited states.

Pooley's next stage requires that the energy is unequally shared between the two ions. Because of the symmetry of the molecule in the lattice the ions separate with momentum vectors along [110] directions. This momentum is then used to initiate a replacement collision chain along a [110] direction. At the starting point of the chain of collision events there will be an F centre, and at the end of the chain the molecular ion will appear as an H centre (i.e. an interstitial halogen). In order to separate this pair of defects to a stable equilibrium, we expect the chain to be four or five stages long. For an alkali halide like KI with a small band gap there is a maximum of 5 eV available, which suggests an energy loss per collision of under 1 electron volt. Various attempts have been made to determine whether this is a realistic energy loss by simulating collisions in a lattice with a computer calculation. The choice of the repulsive potential is important, and in Pooley's first calculations he chose

a Born–Mayer potential, $V(r) = A \exp \left(-\dfrac{r}{B} \right)$, where A and B

are constants. This was encouraging in that it allowed replacement collisions to take place for most alkali halides with the energy available from the exciton, although the length of the collision chain would have been insufficient to achieve a stable separation of the vacancy and interstitial. The harder Born–Mayer–Verwey potential, $V(r) = A + Br^{-12}$, inhibited any collision chains.

Rather than consider a collision series between halogen ions, it is possible to consider that only the smaller halogen neutral is transported from one lattice site to the next. This is probable if the electronic arrangement of the molecular ion allows the charge to be transferred in a different step from the ion collision. Thus successive ions in the chain associate with the next ion, form a molecular ion and move forward to the next site as the neutral ion. Recent calculations[27] support this view. They also indicate that at low temperatures there is a reasonable probability that defocused ions will actually form a metastable pair of closely linked interstitial halogens and an $\alpha(F^+)$ centre.

One advantage of the detailed model by Pooley is that it can be tested by many different types of experiment, and it is encouraging that all the experiments performed so far are compatible with this model. The following list of experiments has helped to credit the model.

(i) There is an anti-correlation between the rates of defect production and luminescence during irradiation. The relative rates vary with temperature and are shown in Fig. 8.13 for measure-

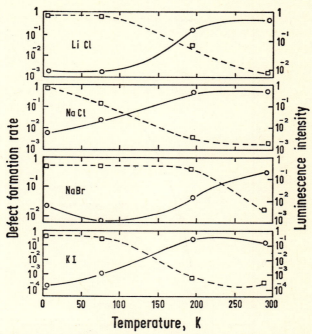

Fig. 8.13 A comparison of F centre production rates ○ and luminescence □ in some alkali halides during electron and X-ray irradiation.

ments made by Pooley[28] for alkali halide crystals. The luminescence was excited by X-rays and the F centre production was with 400 keV electrons.

All the alkali halides show the same general trend, but the magnitude of the effect is different for the different alkali halides. For example the NaCl defect production rate only alters by two orders of magnitude, whereas for KI it changes by four orders of magnitude for the same range of temperature.

The rates predicted by Pooley[24] for the various alkali

halides show these same characteristics, and examples of these theoretical rates are given in Fig. 8.14.

(ii) The threshold energy for intrinsic F centre formation (i.e. not merely charge population of existing vacancies) is less than the band gap energy. For example, new F centres were formed by Goldstein[29] with ultra-violet light irradiations of KI.

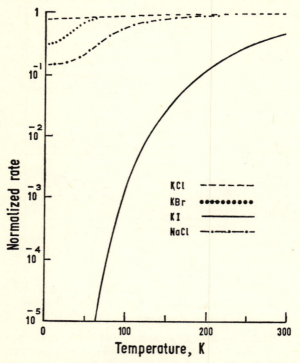

Fig. 8.14 The calculated temperature dependence of the non-radiative recombination rates for electron-hole pairs.

(iii) The pulse experiments of Ueta *et al.*[30] showed that the F centre is formed directly in the ground state, which is consistent with the Pooley–Hersh models but not the other mechanisms of defect formation.

(iv) In a series of mixed crystal experiments by Pooley and Still[31], it was found that the rate of coloration was reduced in mixtures of KCl : KBr and KCl : RbCl. In either case one would significantly disrupt the replacement collision chain so that the halogen vacancies and interstitials could not reach an equilibrium separation.

(v) A more obvious demonstration of the involvement of $<110>$ collision chains of halogen ions comes from low-energy sputtering experiments[32]. In these experiments the alkali halides were sputtered by both photon and low-energy electron irradiation. The ejection patterns show a sharp peak of halogen in the [110] directions which is to be compared with the random ejection of alkali atoms.

The efficiency of sputtering, that is, the number of atoms ejected per incident electron, appears to be a simple function of the geometry of the ions in the lattice and decreases with the parameter S/D, where S is the space between adjacent halogen ions and D is the diameter of the neutral halogen atom. This result is to be expected for any mechanism involving a replacement chain in the $<110>$ direction of halogen.

(vi) The studies of the surface by electron excitation of the Auger electron spectra give details of the concentrations of the atomic species on the surface, and hence the mechanisms of atomic loss from the surface. One finds that during the measurement, which is also an electron irradiation, the surface dissociates with an initial loss of halogen atoms. The equilibrium surface then depends on the temperature, since this controls the rate at which the metal ions can diffuse or evaporate.

As an example of the study of mechanisms of defect formation the work of F and H centre production in alkali halides is particularly instructive, well developed and corroborated by a diversity of experiments. The same cannot be said for any other defect centres. For example, the photographic process already mentioned in § 3.7, despite its immense value and applications, is far less well understood. In fact it is a tribute to the patience and intuition of photographic workers that so much progress has been made in the preparation and sensitization of emulsions without the aid of an identified model for the formation of a latent image. We shall again mention the initial stage of latent image formation, and for details of other processes in photography the reader might consult the book by Mees and James[33] or the article by Slifkin[34]. There is some similarity with the alkali halides in that the final defect is formed by a low-energy photon (such as an exciton) and halogen is released from the surface of the silver halide grain. However, the material is extremely complex, because the actual latent image depends on the history of the grains, the purity, temperature, light intensity and the material of the matrix in which the grains are embedded. We thus guess that electron and hole trapping as well as atomic displacements are important. The flux dependence (reciprocity effect) also suggests that this is at least a two-stage process. The early theory of Gurney and Mott[35] envisaged the light being absorbed to produce an exciton which was then dissociated into an electron-hole pair. They assumed that in the lattice were interstitial atoms, either silver or some unspecified impurity. This trapped the electron to become an Ag^- ion, while the hole was conveniently lost at some hole trap. There would be a coulombic attraction between the charged interstitial and the positive silver ions of the lattice which brought the two together to form a neutral silver molecule. This unit could then act as the nucleation site for the growth of a colloidal speck of silver. The flux dependence would follow if the initial Ag_2 were unstable and Ag_3 or a larger speck were needed for stability. Halogen escape from the surface to the gelatin could occur by neutralization of the surface halogen atoms by the hole.

A later theory by Mitchell and Mott[36] avoided the problem, that at room temperature the number of interstitials in the silver halide was negligible, by trapping the hole at a kink site or a surface and forming a free positive silver ion. This is then assumed to diffuse through the lattice as a defect capable of being pinned by another imperfection. If it captures the photoelectron, the silver becomes neutral and again forms the first stage in the development of a silver speck.

Despite a wealth of experimental details which may be compatible with these theories there are no positive tests which finalize the specific model. This is not surprising considering the speculative nature of the 'special sites' required for the latent image to form and the fact that electronic motion is the major step in either model. However, electronic motion to separate the electrons and holes will proceed under the action of the electric field which extends from the surface into the bulk[37].

8.13 Summary

In this chapter we have shown that both intrinsic defects and impurities may be introduced into materials by a variety of means and we are not necessarily limited to conventional chemical additions to the melt before growth. It is also apparent that the mechanisms by which defects result from low-energy irradiation are not well understood, except for the alkali halides where the Pooley–Hersh model seems to be substantiated. In many other substances which colour with small amounts of energy, for example, photographic grains MgF_2 and sodalite, there is still work to be done to determine the details of the processes involved.

Chapter 8 References

[1] Yahia, J., *Phys. Rev.* **130**, 1711, 1963.

[2] Dearnaley, G., *Rep. Prog. Phys.* **32**, 495, 1969.

[3] Bayly, A. R. and Townsend, P. D., *Optics and Laser Tech.* **2**, 117, 1970.

[4] Wilson, R. E., *Optics and Laser Tech.* **2**, 19, 1970.

[5] Thompson, M. W., *Defects and Radiation Damage in Metals* (Cambridge) 1969.

[6] Chadderton, L. T., *Radiation Damage in Crystals* (Methuen) 1965.

[7] Carter, G. and Colligon, J. S., *Ion Bombardment of Solids* (Heineman) 1968.

[8] Lindhard, J. and Scharff, M., *Phys. Rev.* **124**, 128, 1961.

[9] Piercy, G. R., Brown, F., Davies, J. A., and McCargo, M., *Phys. Rev. Letters* **10**, 399, 1963.

[10] Mott, N. F., *Proc. Roy. Soc.* **A124**, 426, 1929; ibid. **A135**, 429, 1932.

[11] McKinley, W. A. and Feshbach, H., *Phys. Rev.* **74**, 1759, 1948.

[12] Corbett, J. W., Supp. **7**, *Solid State Physics*, 1966.

[13] Glendennin, L. E., *Nucleonics* **2**, 12, 1948.

[14] Katz, R. and Penfold, A. S., *Rev. Mod. Phys.* **24**, 28, 1952.

[15] Kobetisch, E. J. and Katz, R., *Phys. Rev.* **170**, 391, 1968.

[16] Erginsoy, C., Vineyard, G. H. and Englert, A., *Phys. Rev.* **133**, A595, 1964.

[17] Lomer, J. N. and Pepper, M., *Phil. Mag.* **16**, 1169, 1967.

[18] Crawford, J. H., *Advances in Physics* **17**, 93, 1968.

[19] Seitz, F., *Rev. Mod. Phys.* **26**, 1, 1954.

[20] Varley, J. H. O., *Nature* **174**, 886, 1954.

[21] Dexter, D. L., *Phys. Rev.* **118**, 934, 1960.

[22] Williams, F. E., *Phys. Rev.* **126**, 70, 1962.

[23] Klick, C. C., *Phys. Rev.* **120**, 670, 1960.

[24] Pooley, D., *Solid St. Comm.* **3**, 241, 1965; *Proc. Phys. Soc.* **87**, 245, 1966.

[25] Hersh, H. N., *Phys. Rev.* **148**, 928, 1966.

[26] Murray, R. E. and Keller, F. J., *Phys. Rev.* **137**, A942, 1965.

[27] Smoluchowski, R., Lazareth, O. W., Hatcher, R. D. and Dienes, G. J., *Phys. Rev. Letters* **27**, 1288, 1971.

[28] Pooley, D. and Runciman, W. A., *J. Phys. C.* **3**, 1815, 1970.

[29] Goldstein, F. T., *Phys. State. Sol.* **20**, 379, 1967.

[30] Ueta, M., Kondo, Y., Hirai, M. and Yoshinary, T., *J. Phys. Soc. Japan* **26**, 1000, 1969.

[31] Still, P. B. and Pooley, D., *Phys. Stat. Sol.* **32**, K147, 1969.

[32] Elliott, D. J. and Townsend, P. D., *Phil. Mag.* **23**, 249, 1971.

[33] Mees, C. E. K. and James, T. H., *The Theory of the Photographic Process* (Macmillan N.Y.) 1966.

[34] Slifkin, L., article 'The Photographic Process' to be published in *Solid State Physics*.

[35] Gurney, R. W. and Mott, N. F., *Proc. Roy. Soc.* **A164**, 151, 1938.

[36] Mitchell, J. W. and Mott, N. F., *Proc. Roy. Soc.* **8**, 1149, 1957.

[37] Kliewer, K. L., *J. Phys. Chem. Solids* **27**, 705, 1966.

9

APPLICATIONS OF DEFECTS
TO PRACTICAL DEVICES

9.1 Applications of defects to practical devices

In this chapter we give some selected examples of the role of defects in insulators and semiconductors to show how they control the performance of commercial devices. The field from which we can choose examples is enormous, because all the major effects of semiconductor physics rely on the control of the Fermi level and charge mobility through various impurities or defects. Similarly solid state lasers and luminescent materials operate with a host lattice which has rare earth or other impurity ions dispersed in the matrix. Despite this enormous range of potential examples we all too often find that the detailed mechanism of control, or even the defect site, is not known. In practice the Fermi level may be controlled empirically by including into the sample such a high concentration of a single dopant that the effects from all the other impurities and defects are minimized. The material may then show the desired properties even though we are ignorant of the detail of the defect sites and their interactions. This type of control is only a first order stage of development, because to specify the electron or hole mobilities and their temperature dependence requires a more sophisticated control of *all* the impurity levels. This is particularly true when one is concerned with defect concentrations of only parts per million (p.p.m.) of the lattice sites, for this is comparable with the concentration of 'unintentional' impurities. There is also the problem of knowing whether the impurities are substitutional, interstitial, aggregated or localized near grain boundaries.

In luminescent materials, both photon- and electron-stimulated, this situation frequently exists, and it is still true that many phosphors are prepared by a standard recipe and a little prayer that the source material contains the correct ratio of impurities. This faith is not always justified, for as described in Chapter 4 the more sensitive techniques, such as thermoluminescence, show large variations in properties even from different sections of the same crystal. The emotive words used in phosphor preparation are activator, co-activator, network former, inhibitor, poison or killer, and such mysticism exists because we generally cannot quantitatively state that 'X per cent of impurities A sit in substitutional sites next to vacancies of type B'. On the basis of work in the alkali halides and the rate of progress in identification it appears likely that intuition and 'magic brews' will play a role in this field for many years to come.

The problems are not insoluble, and where there is sufficient commercial impetus, for example in the purification of germanium and silicon, suitable methods have been devised. We should not deny the value of the intuition gained by experience, because many devices will operate without a full understanding of the important defect sites. It is only when we attempt to optimize materials that a deeper understanding is necessary. As examples we will give an alphabetical list of some of the processes that are controlled by defects. We shall follow this with more detailed discussions for some items, for others only one or two references. The literature of semiconductors and luminescent materials is very large, and our references are intended to be typical examples and not an exhaustive list of the field.

Chemical analysis. Dark trace image tubes. Dating of pottery. Effects of radiation damage on semiconductors. Electroluminescence. Electro-optic devices. Information storage. Laser glasses. Luminescent materials. Mechanical strength. Optical filters. Optical materials. Particle detectors. Passivation of surfaces. Photochromism. Photoconductors. Photo-etching materials. Photography. Photomultiplier cathodes. Purification of materials. Radiation dosimeters. Reactor fuel elements and materials. Secondary electron conduction. Semiconductors. Thermally stimulated colour changes.

9.2 Defects and semiconductor applications

Conduction in semiconductors can obviously result from the promotion of electrons from the valence band to the conduction band by thermal or optical excitation. At room temperature ($kT = 0{\cdot}026$ eV) the band gaps of silicon or germanium ($1{\cdot}16$ or $0{\cdot}74$ eV) produce intrinsic conductivities of only 10^{-3} ohm^{-1} cm^{-1}, whereas doped samples may be used with conductivities of 10^{+2} ohm^{-1} cm^{-1}. Optical excitation will produce photoconduction by band to band transitions, and the intrinsic material may be used as a photodetector. However, to maintain control over the conductivity at room temperature it is simplest to trap electrons at the shallow levels produced by impurities. We expect that the group IV elements will accommodate the similar size ions from groups III or V as substitutional impurities, and the extra electron or hole (i.e. electron deficiency) of these ions will produce electron or hole (n or p) type conductivity. This is indeed so, and up to 0·1% of the lattice sites can be filled with impurity ions without serious lattice distortion. The binding energy, or trap depth, of the electron or hole is indicated in Table 9.1 for some typical impurities in germanium and silicon. Such shallow traps are emptied at room temperature.

Table 9.1 Trap depths of e^- or h^+ in Ge and Si in eV

	n-type impurity			p-type impurity		
	P	As	Sb	B	Al	Ga
Ge	0·0120	0·0127	0·0096	0·0104	0·0102	0·0108
Si	0·045	0·059	0·039	0·045	0·057	0·065

From a defect viewpoint it is immaterial whether the impurity ions are introduced during crystal growth, by diffusion or by ion beam implantation, assuming that no other defects are produced at the same time. Also for our purposes transistor technology is now 'just' a matter of making suitable junctions with p, n or intrinsic regions and attaching these to electrical contacts.

In semiconductors, as in insulators, photoconductive phenomena can be limited to some specific spectral region by choosing a suitable defect level and maintaining the crystal at a temperature at which the electron is not thermally excited. An excellent introduction to this field is provided by the book by Bube[1]. An example of a photoresponse spectrum is given in Fig. 9.1 for CdS doped with copper or silver.

Photo-ionization also means that semiconductors are suitable materials for detection of radiation[2] such as gamma rays, electrons and alphas, since the ionization losses produce electron-hole pairs which can be electrically detected. To separate the charges and produce signal gain the junction region of a transistor is ideal. Not only can one determine the number of pulses received, but also the energy of the particle can be obtained by the total number of electron-hole pairs that are liberated.

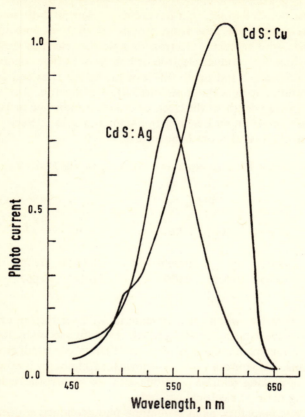

Fig. 9.1 The photoconductive response of CdS doped with copper or silver.

To obtain a large sensitive region it is common to use a detector containing lithium impurity. This is interesting because unlike the group III and V elements it does not occupy a substitutional site on the lattice but instead is an interstitial. Also because it is a small ion in the silicon lattice it can diffuse very readily.

In practice one diffuses lithium thermally into the semiconductor and then drifts the ions under a reverse bias to produce a uniform plateau of acceptor levels over several millimetres of the sample. The control of this impurity profile gives excellent particle detection characteristics, and the device is called a lithium-drifted detector.

We also should note that radiation damage can be produced both in particle detectors or other transistors. Because of ionization,

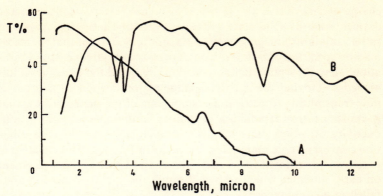

Fig. 9.2 The transmission spectra of *n*-type Si measured at 18 K. Curve A was the original spectrum, B was after a neutron dose of 6×10^{17}nvt.

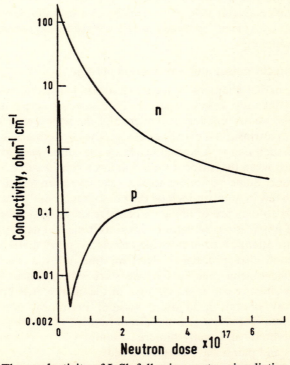

Fig. 9.3 The conductivity of InSb following neutron irradiation.

charge exchange and the production of new levels, semiconductor devices, such as solid state radiation detectors and transistors, can suffer radiation damage. For example, in Fig. 9.2 we see that the transmission spectrum of n-type silicon changes totally after neutron irradiation. New absorption bands have appeared, and the Fermi level has moved so far that the initially opaque region beyond 10 microns is now transmitting. Electrical measurements of the damage show that a similar neutron irradiation of indium antimonide lowers the conductivity of n-type material. More dramatically, prolonged irradiation converts p-type material to n-type, as in Fig. 9.3. This behaviour is serious for transistors in radiation environments, and equipment on early satellites which passed through the earth's radiation belts failed as a consequence.

Impurity defects need not play an active electrical role in transistors and indeed can be deliberately added to a surface layer to produce chemical inertness. The silicon-oxygen reaction to form inert, insulating SiO_2 is an obvious example. The modern MOSFET (metal oxide semiconductor field effect transistor) uses a thin layer of SiO_2 as an insulating layer between the field electrode and the bulk of the semiconductor.

9.3 Photochromism and information storage

In the earlier chapters we have placed particular emphasis on the study of intrinsic defects, because if we are unable to understand these then we are unlikely to understand the situation in which we have both intrinsic and extrinsic defects. The strength of the chemical bonds is such that in a normal insulator the colour centres will only be formed when the incident radiation has a considerable amount of momentum. Some exceptions to this rule have been mentioned and, although few in number, include the photographic and photochromic materials so they are of immense commercial interest. For the purposes of applications of colour centres to practical devices we must move our attention from intrinsic defects in 'pure' materials to impurity-controlled processes. Here we find all the usual colour centre phenomena, but because we start with charges trapped at impurity sites we can make changes in the charge distribution with low-energy photon irradiation, without necessarily creating new defects. Sunlight contains a significant number of high-energy ultraviolet photons, and many hundred examples of colour changes in minerals and paints have been recorded.

If the changes result from charge exchange, the process is usually reversible by a further optical or thermal bleach, and such materials which colour with one wavelength and bleach with another are termed photochromic. For the purpose of our discussion we shall include materials which colour with electron irradiation, although the appropriate name should be cathodochromic.

The three major steps in a photochromic cycle are the production of an optical absorption band, detection of the change and a reversal

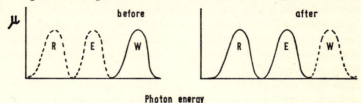

Fig. 9.4 The absorption spectrum before and after writing in a photo-chromic material. The bands are used for, R, reading, E, erasing, and W, writing.

process. Since the usual objective is information storage these are called write, read and erase stages. An ideal photochromic material might show absorption spectra like that of Fig. 9.4. Absorption of light in band W will reduce band W and produce two further absorption bands R and E. These bands differ in that band R is stable against optical bleaching, whereas photon energy absorbed by band E will

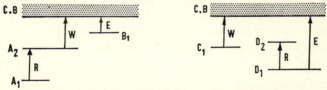

Fig. 9.5 Two energy level schemes which produce photochromism.

cause the bleaching of both the R and E bands and a simultaneous reappearance of band W. The three processes of writing, monitoring and erasing are completely separate in this example and could be achieved in several ways. The energy level diagrams of Fig. 9.5 show two suitable examples. In the first case the levels A_1 and A_2 relate to a single defect or impurity, and level B_1 is the electron level for a different impurity. If both A_1 and A_2 are initially populated, then the transition R is forbidden. However, when an electron is stimulated from A_2 by a photon of energy W, the electron will be transferred to level B_1. Absorption of light in band R is now possible, so the strength of this band monitors the change in population of the level A_2. The states A_1, A_2 might correspond to a multiple-charged ion in the host lattice or a rare earth ion which has internal transitions. The erase step corresponds to a bleaching of the shallow level B_1, which can, in this example, proceed by either an optical or a thermal bleach.

A second method of producing these three absorption bands is to start with a filled level C_1 and empty levels D_1 and D_2. Absorption in the band W transfers charge to the ground state D_1. Excitation from D_1 to the higher bound state D_2 gives a monitoring band R and de-excitation of the system follows by transitions to the conduction band by an energy E.

With such idealized systems all three bands are separate, there are no other competing traps or back reactions, and bands W and E are totally bleachable. In practice bands R and E could be the same, if only a low-intensity light was required for monitoring or if the stored information was to be read only once.

Photochromism is not limited to inorganic insulators, and precisely the same effects are achieved by chemical changes produced in organic materials[3, 4, 5]. We may choose to describe the processes in terms of chemical compounds, but so long as only charge transfer is involved in the reaction there is the possibility of reversing the process. Organic systems have potentially more storage capacity than doped insulators, because each molecule can be a storage element, thus high optical storage densities are achieved and even thin tapes of material will record intense coloration. A recent review paper by Jackson[3] lists the common organic photochromic classes as:

(i) Trans-cis isomerism caused by rotation about a double bond (e.g. azo compounds).
(ii) Bond rupture (e.g. spiropyrans).
(iii) Transfer of an H atom to different positions in a molecule (e.g. anils).
(iv) Photo-ionization (e.g. triphenyl methane dyes).
(v) Oxidation-reduction reactions (e.g. thiazine dyes).
(vi) Excitation to metastable triplet states.

To apply the photochromic materials to specific problems we need to consider the ease of coloration, efficiency and speed, and the stability of the centres so produced. Both electron beams and ultra-violet light sources contain sufficient energy. An electron beam can easily be directed and modulated, but requires a vacuum chamber and may also excite other levels than those intended. Optical beams need high power levels, which is a disadvantage unless we operate at a laser frequency, and these are less readily controlled spatially, which is necessary in scanning through memory locations. However, for image storage this is unimportant. Similarly the speed of writing depends on the application, for example in a window glass which darkens in sunlight the change may take an hour, but for a high-speed computer tape the information must be written in the stable form in some nanoseconds. That is, not only must the electron be transferred, but also the 'read' centre must reach the ground state. This excludes many organic systems which pass through excited states and hence have lifetimes of milliseconds for the charge transfer.

With photochromic materials there is a problem of producing a sufficient concentration of defects to give a strong visual absorption band, because only one electron is transferred per incident photon. To achieve an optical density of, say, unity, that is a transmission of 10%, in a 1 mm sample we need an absorption coefficient of 23 cm^{-1}, which we can relate to the defect concentration by Smakula's expression (given in Chapter 3 as $\mu_{max} W \approx 10^{-17} Nf$). For our 1 mm sample this typically means N is 10^{18} centres per cm^3. So the incident light beam must contain at least 10^{18} photons. Our writing speed thus depends on the light intensity available, but fortunately fluxes of 10^{25} photons cm^{-2}s^{-1} are obtainable from laser sources. Tests made with sodalite[6] have produced colorations of 0·4 optical density following a single laser pulse of 10 nanoseconds' duration. With this time scale computer applications of photochromics are feasible. The final erase step may be achieved by broader band illumination or heating.

We have already discussed the role of silver halides in the photographic process, so we might expect that they could act as photochromic materials if held in a suitable matrix. Glass matrices[7, 8] might typically contain oxides of SiO_2 (60%), Na_2O (10%), Al_2O_3 (10%), and to this B_2O_3 (20%) is added as a flux. Photochromism is induced by traces of silver halide and a sensitizer such as CuO. The similarity between these glasses and photographic emulsions occurs, because the silver halide phase separates out from the matrix and crystalline silver chloride can be identified. The separation is possible, because the melting point of the silver chloride is below the strain annealing temperature range of the glass. The size of the dispersed spheres of silver halide is controlled by the annealing cycle of the glass. In a clear glass the spheres are 5 to 10 nm in diameter. Larger spheres, say 30 nm diameter, produce light scattering and the glass is opalescent. The sensitive spheres are typically dispersed at an average distance of 60 nm, and since this is much less than the wavelength of visible light or the resolution of an optical system we can record all the image detail that is required. The magnitude of the absorption within each island of AgCl depends linearly on the amount of irradiation it receives. The similarity to the photographic process probably extends to an image-forming reaction of

$$AgCl + h\nu \rightarrow Ag^\circ + Cl^\circ + e^- + h^-$$

as is evidenced by the wavelength and temperature dependence of the excitation spectrum. However, the difference in matrix, glass or gelatin is very important, as the halogen ion cannot escape so the process is thermally or optically reversible. Examples of the absorption spectrum of the glass before and after exposure to light is shown in Fig. 9.6, and the temperature dependence of the rate of

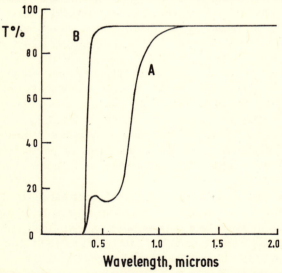

Fig. 9.6 The transmission spectrum of a photochromic glass after irradiation A, and after bleaching B.

darkening and fading of a glass is indicated for a Corning glass in Fig. 9.7.

When the glass is irradiated with polychromatic light, there is a dynamic balance between bleaching and coloration which poses some

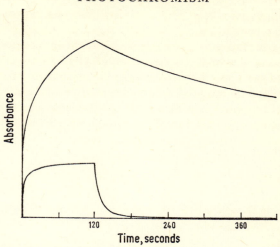

Fig. 9.7 The colouring and bleaching curves for two glasses. Irradiation
         time was 120 seconds.

problems in the choice of band positions. For applications such as
self-adjusting sunglasses or sun-excluding windows there is the problem that the formation band is in the ultra-violet whereas sunlight
consists of predominantly lower energy photons which produce
bleaching. The performance of two window glasses is recorded
in Fig. 9.8 during the course of 24 hours. As the light intensity and
the amount of ultra-violet increases the window becomes darker,
but it clears again at sunset. The speed of response and the final

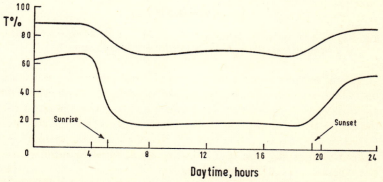

Fig. 9.8 The transmission of two window glasses throughout a period of
         24 hours.

equilibrium levels differ for the two nights, because there was a temperature change.

For our second example we will choose more 'classical' photo-chromic examples of CaF_2 doped with CeF_2 or CeO_2. These are classical in the sense that only charge exchange is involved and no atomic diffusion is required. Cerium shows some of the characteristic features of rare earth ions by producing broadened line spectra after excitation. The spectrum for $CaF_2 : CeF_2$ in Fig. 9.9 was obtained at

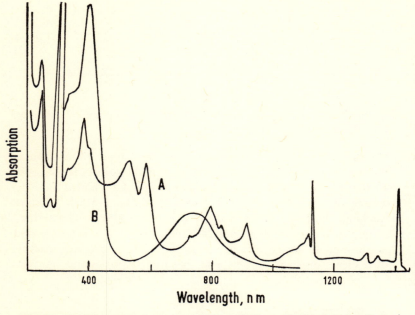

Fig. 9.9 The absorption curve of $CaF_2 : CeF_2$ at 78 K after irradiation with ultra-violet light A, and in the bleached state B.

78 K and initially gives a typical broad band picture for defects in insulators, but when excited with ultra-violet light the line spectrum appears. Reversal is possible for bleaching light above 450 nm. From the shape of the absorption features one suspects that the cerium is linked to an intrinsic CaF_2 defect[9]. The $CaF_2 : CeO_2$ system[10] is closer to our model case with a visible region which is initially clear, but when one excites with high-energy photons ($\lambda < 250$ nm) two new absorption bands appear. The spectrum, in Fig. 9.10(a), shows they are well resolved and they have the properties that the 580 nm

band is stable, but light absorbed in the 380 nm band produces bleaching of the entire system back to the initial state.

Many photochromic materials do not produce such well defined optical absorption bands but only an overall increase in the absorption spectrum. The example of $SrTiO_3$ doped with Fe and Mo or Ni and Mo is well documented[11], and it is known that the impurity ions Fe^{3+} and Mo^{6+} occupy Ti^{4+} sites, and charge equilibrium is

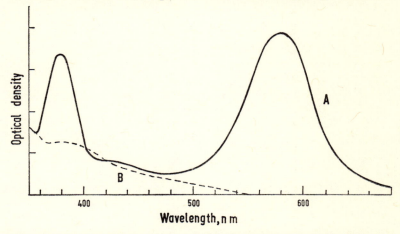

Fig. 9.10(a) Absorption curves of photochromic $CaF_2 : CeO_2$. Curve A is produced by ultra-violet light, B by bleaching.

maintained if there are twice as many iron ions as molybdenum ions. Neither ion shows visible absorption bands, but when stimulated with ultra-violet light in the range 390–430 nm the visible region becomes opaque with colorations up to 100 cm^{-1} caused by the transfer of electrons to form Fe^{4+} and Mo^{5+} ions. The system is stable at 78 K but has a half-life of minutes at 300 K. Reversal is possible with light from 500 to 800 nm, as may be expected from the absorption curve in Fig. 9.10(b). The intense coloration is possible, because we require only charge exchange, and we can readily replace 1 in 10^4 of the titanium ions with impurities, which gives an effective defect concentration of 10^{18} cm^{-3}. This strontium titanate system has been used[12] to store two-dimensional arrays of information up to 10^4 bits cm^{-2}, and it withstood up to 10^7 readout cycles.

In holographic storage one measures the efficiency by comparing the intensity of the reconstructed light beam with the reference beam used for the readout. In the case of storage by absorbing centres there is a maximum theoretical efficiency of 3·7%, and this figure is approached in the $SrTiO_3$ case where efficiencies of 1·2% have been recorded. It is worth noting that this is better than that normally

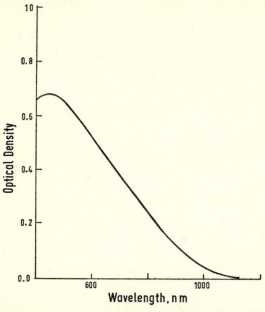

Fig. 9.10(b) Absorption curve of $SrTiO_3$ doped with Fe and Mo.

possible in photographic plates. Amodei and Bosomworth[12] also used the $SrTiO_3$ system to store a hologram by starting with the coloured state and bleaching in the signal with a laser pulse. Using a ruby laser they wrote the hologram in less than 20 nanoseconds.

9.4 Other methods of information storage

One problem of using absorbing centres to diffract a laser beam for holography is the maximum efficiency of 3·7% for the light intensity emerging in the reconstructed beam. In a non-absorbing system the efficiency could approach 100%. To take advantage of this fact we can store the information by changes in the refractive index. This is possible in crystals of lithium niobate or lithium tantalate[13], where non-linear optical effects occur at high light intensities. The changes in the crystal are permanent, and the resultant internal electric fields modulate the refractive index in the desired fashion to a depth of 1 part in 10^3 in the index. Because no absorption bands are involved, the modified crystal cannot be optically bleached and one must anneal it up to 170°C in the case of $LiNbO_3$.

Thaxter (quoted by Taylor)[5] avoided this difficulty with a crystal of strontium barium niobate, where he found that the refractive index change occurred only in the presence of an external electric field. Recycling the sample in this situation requires one to alter this external field.

A totally different approach to information storage via colour centres was proposed by Schneider et al.[14] in which they suggested using alkali halide M centres. These centres have $<110>$ symmetry and can be reoriented without bleaching in a limited temperature range. Thus one can read or write with polarized light beams and erase with unpolarized light. The suggestion is attractive, because it is possible to have some 10^{18} M centres cm^{-3}, but the centres must be maintained below 300 K, and other defect reactions converting the M to F or R centres are possible.

Photochromism is still in its infancy and there will clearly be many more applications. Even now it is possible to obtain self-darkening glass or dolls which 'suntan'. The more sophisticated examples of information storage have such commercial value that much of the work has not yet reached the open literature. To indicate the storage density already achieved we can quote an entire bible printed on a 5 cm square of glass and various holographically written diffraction gratings with line densities of 2100 lines mm^{-1}. Extrapolating this to three-dimensional arrays suggests that storage densities of 10^{10} bits mm^{-3} are possible. However, whether or not one can fabricate a suitable access system is a separate problem. Other suggestions include the transfer of holograms by a television system to produce a T.V. screen which is a replica hologram. This could then be reconstructed by a coherent light beam to give a three-dimensional T.V. image.

9.5 Dark trace image tubes

Normal cathode ray tubes have a dark screen and the electron beam produces a fluorescent spot. One disadvantage of such a display is that the viewing contrast is influenced by the ambient light level and only poor contrast is seen in normal room light. The alternative system is to write a black beam on a white screen, then the contrast is enhanced by room light, as with book printing. Cathodochromic materials offer the possibility of making such a dark trace screen, and this was suggested as early as 1940[15]. Work in this area[16] showed that the F centres in KCl evaporated films could be used to produce the images on the screen. However, the fade times, complex reactions to form M or R centres and short operating life excluded the material for most purposes. One application in which the long fade time was particularly useful was the presentation of radar displays. As the radar antenna scans a region the echo signal is shown as a spot on the screen. If the object is moving then the spot also moves, and it is desirable to show not only the object's position, but also its direction and speed. A photochromic screen with a slow fade time is ideal for this purpose, because the radar spots record coloration along the aeroplane's path, but the fading means that only some 20 seconds of path are recorded at any one moment and the early part of the path is faded compared with the latter part. Hence the length of coloration gives a vector representation of the speed, and the faded tail shows the direction of the motion.

More modern tubes use sodalite rather than KCl, because sodalite does not form complex centres so readily and it can be bleached by room light or an occasional rapid heat or optical bleach. Various dopants are currently being tried to extend the operating life of the tube beyond a few hundred hours or to alter the fading characteristics.

In a photochromic material the host lattice or glass matrix is chemically altered, but so far we have only discussed this new state in electronic terms and examined the optical properties. However, the differences between the perfect and altered material can also be revealed by a chemical reaction[17, 18], hence we can etch a pattern into the glass on a scale determined by the resolution of the optics used for image production. One such example is a silver nucleated opal glass which under illumination develops a crystalline phase of lithium metasilicate. This is highly soluble in dilute hydrofluoric acid and can be selectively etched from the glass without noticeable attack occuring on the rest of the glass. By this technique glass screens have been made up to 300 mesh and systems of fluid amplifiers have been chemically machined from the glass.

With this particular glass there is also the possibility of heating the finished product to convert the glass to lithium disilicate, which is a hard, ceramic-like material.

9.6 Optical materials

When choosing optical materials there is invariably a conflict between the properties that one wants and the available compounds.

The competing requirements include the spectral range for which the material is transparent and thermal, mechanical, radiation or chemical stability. The items must also be compatible with other components; for example, in a system with a window bonded on to a container the thermal expansion coefficients and chemical reactions must be considered. Severe problems arise if the window must be a large single crystal which is free of strain, and the converse problem in amorphous materials is to avoid the formation of crystallites.

The standard optical problem is to transmit or attenuate a particular set of wavelengths, and depending on the spread of energies involved we describe this as a window or a filter. In the case of windows the intrinsic absorption edge sets an upper energy limit for the transmission, and the characteristics of the rest of the spectrum will depend on defect absorption or lattice vibrational absorption bands. To improve the overall absorption we must suppress the defect bands either by removing the defects or by adding alternative impurities, so that the Fermi level is altered and absorption bands in one region of the spectrum disappear, even at the expense of producing other bands. For compound materials purification is difficult, so the doping method is generally used. In wide band gap materials this approach is used in silica, rutile and lithium fluoride. Silica is important because the glass is rugged, chemically stable and can be used with graded seals on to other glassware such as photomultiplier tubes. To utilize the transmission of air through the visible and ultra-violet region (1 to 6 eV) we must remove the strong silica absorption band which occurs below 240 nm.

If an additonal electron trap is introduced into the crystal in the form of water the band is suppressed by charge exchange or compensation. Unfortunately the infra-red spectrum now shows the characteristic vibrational absorption band of the OH radical at 2·8 microns. This is an acceptable compromise, since there is a greater choice of narrow band gap materials which may be used as alternatives for the long wavelengths measurements. Fig. 9.11 shows examples of the ultra-violet and infra-red spectra of the water-free and doped silica.

Even if one uses a dopant that does not itself have absorption features, we may find infra-red absorption from lattice phonons which were previously forbidden, but whose selection rules are relaxed in the imperfect lattice.

In the vacuum ultra-violet the ultimate limit of window materials is set by lithium fluoride with a band gap of 12·9 eV. Here we have a particularly complex defect problem, because not only do impurity

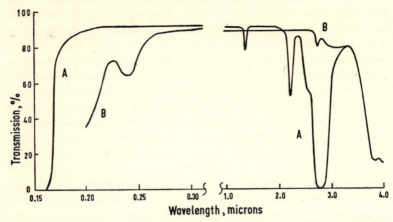

Fig. 9.11 The ultra-violet and infra-red spectra of silica, containing water A, and water-free B.

bands occur near the band edge, but also absorption of light at the edge produces new defects in the alkali halides (see Chapter 8). Purity control will help with the first problem, but the intrinsic defect formation can only be suppressed by doping. The problem is not yet solved.

At lower energies there is a wide selection of band gaps and refractive indices, and many glass filters are available either as cut-off

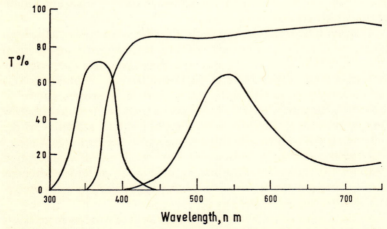

Fig. 9.12 Three examples of the transmission curves of commercial glass filters.

filters (e.g. photographic haze filters) or selective energy filters (e.g. heat-absorbing filters). Most glass companies offer a range of such filters, and some typical spectra are shown in Fig. 9.12. Filters with high-energy transmission and visible absorption cannot be made with such flexibility in design except by reducing the overall transmission

In the infra-red region the problem is not the size of the band gap but the low-energy end of the spectrum where lattice absorption occurs. The transmission limits for a range of infra-red materials are given in Table 9.2. The basic constituents of the Irtran series are also quoted; however, the materials are formed under pressure at high temperature and are therefore not crystalline and do not cleave. In some cases the OH bands at $2 \cdot 8$ to $3 \cdot 0 \, \mu$ (2800 to 3000 nm) can be removed by subsequent vacuum reheating.

Table 9.2

	Material	Transmission limit (microns)
	SiO_2	$4 \cdot 5$
Irtran 1	MgF_2	$9 \cdot 0$
Irtran 5	MgO	$9 \cdot 5$
Irtran 3	CaF_2	$11 \cdot 5$
	BaF_2/caF_2	$12 \cdot 0$
	Ge	$25 \cdot 0$
	$NaCl$	$26 \cdot 0$
KRS5	$TlBr/TlI$	$40 \cdot 0$

9.7 Luminescent devices

Under this category we include phosphors[19], solid state lasers[20], electroluminescent materials[21] and the scintillator type of particle detector. The defect roles are obvious in that energy is absorbed in the crystal either by electron-hole pair production or by charge excitation at some localized site. The excited crystal then decays by a radiative process to the ground state, either emitting energetic photons by a direct band-to-band recombination or, more normally, decaying to a recombination centre. The problem is to choose the suitable defects for the excitation and recombination centres. For visible emission one needs wide gap materials and deep impurity levels or internal transitions as in the rare earth ions. Although intrinsic defects alone might be suitable, all commercial phosphors include impurity ions, some of which produce charge compensation for intrinsic defects, others associate into complexes with intrinsic defects and others merely act as luminescent sites.

Lasers are just a special case of normal phosphors which give narrow line emission. They may need more care in preparation to avoid unwanted impurities, which absorb energy from either the exciting or emitted light. To achieve sufficient light intensity for laser action also means a very high concentration of defects and good single crystals. So laser materials are limited to those substances that can accommodate impurity ions without gross distortion. The first laser glass, ruby, is a good example of this, for the chromium is accommodated on the aluminium site of the Al_2O_3 up to at least 1% of the lattice sites, without excessive strain.

9.8 Thermoluminescent dosimetry

In the discussion of thermoluminescence in Chapter 4 we showed that the technique is extremely sensitive and might detect as few as 10^9 centres in a specimen. This sensitivity is valuable even if the total imperfection content of a sample is much higher, because the measurement separates the different trapping levels into separate glow peaks. These two properties, sensitivity and discrimination, make glow curve measurements ideal for radiation dosimetry. We merely require that there is a suitable empty trapping level in a material, which is populated during irradiation, and that the fraction of these levels which become filled can then be read out by thermoluminescence. The design specifications for the material should include the following:

(i) A stable wide band gap insulator.
(ii) A controlled concentration of deep electron trapping sites which are initially empty.
(iii) Trap filling should depend linearly on the total irradiation dose and be independent of the rate of ionization in the material or fluctuations in the ambient temperature.
(iv) A controlled concentration of recombination centres.
(v) A negligible concentration of alternative trapping or recombination sites.
(vi) Preferably a material that can be thermally recycled and used many times.
(vii) For personnel dosimetry we also require the cross sections for radiation absorption to be comparable with that of a human body.
(viii) Finally, we need a simple but reliable readout system, and the

overall cost of the system must be competitive with conventional film badge dosimetry.

In practice the system operates with a powder sample which records the irradiation dosage. The stored energy is then read out by rapidly heating the powder in a hot dish at, say, $20°C \ s^{-1}$, and detecting the light with a photomultiplier tube. To improve the sensitivity and reject the thermal radiation it is customary to use a blue transmitting filter. Various signal display modes are possible and include the total emission curve, peak height displays or the area beneath the glow curve over a particular temperature range. Depending on the signal strength one can use the phototube in a normal d.c. current mode or for very low light levels in a pulse-counting mode. We are not concerned with the details of the apparatus, but a commercial readout unit can be suitable for semiskilled operators without any loss in reliability.

When considering our generalized specification for the dosimeter powder we attempted to avoid the shortcomings of the X-ray film badges to which the powder is an alternative. In particular we would like a wider range of signal storage without the equivalent of a reciprocity failure at low fluxes or film saturation at high fluxes. The simplicity of the thermoluminescence readout also avoids the errors introduced in the development of the film processing and of course the powder is re-usable.

Commercial powders are available that approach our specifications, and the major material is LiF doped with magnesium and minor trace elements. Other materials which have been used are calcium sulphate with manganese or lithium borate glasses. The precise defects involved and the role of the trace impurities is only partially understood, and a great deal of the sensitization and doping technique is still in the realm of a magic art. We will now attempt to explain the purpose of some of this art.

Lithium fluoride is a suitable host material with a wide band gap, stable chemistry and a suitable radiation cross section for personnel dosimetry. Purification is difficult, so rather than use an intrinsic defect level the material is heavily doped with magnesium to outweigh the other trapping levels. Magnesium produces five glow peaks in the LiF[22] at temperatures from 80 to 225°C (see Fig. 9.13). With suitable thermal treatments only the deepest trap gives a significant peak, so the irradiated powder can be stored without emptying the trapping level at ambient temperature.

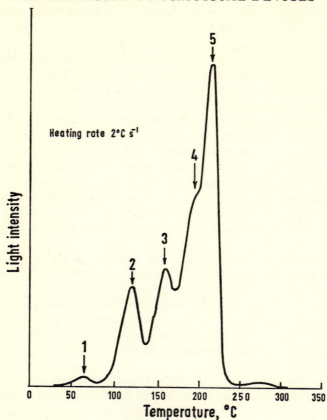

Fig. 9.13 Thermoluminescence in LiF : Mg.

The powder responds linearly to irradiation dose[23] and is in-sensitive to the irradiation flux, as we are only filling electron traps which exist in large concentrations because of the magnesium and should be contrasted with the behaviour of intrinsic F and M centre populations formed by the irradiation (see § 3.7). The radiation response is linear from millirads to thousands of rads, and the powder can be re-used many times without deterioration if given an appro-priate thermal treatment. One may note that for medical applications radiation workers may 'safely' receive dosages of 100 m rad per week. The LiF : Mg dosimetry range extends to much higher levels of 10^4 rads, but here the only significant application would be in a post-mortem.

To sensitize the LiF one adds Mg to, say, 100 p.p.m. and, intentionally or not, other trace elements. The powder is then heated to 400°C for an hour and quenched to room temperature. In this condition the magnesium is dispersed in various states in the lattice, and an irradiation followed by a glow curve gives five peaks as shown in Fig. 9.13.

Parallel measurements[24,25] of the optical absorption bands show (Fig. 9.14) the irradiated material contains bands at 113, 135, 195, 205, 250, 310, 380 nm. Efforts to identify the defects have come from two directions. The first is the standard approach of noting the effects of bleaching with various wavelengths and the changes produced by annealing cycles. These experiments have produced useful correlations which show that the glow peaks labelled 4 and 5 both

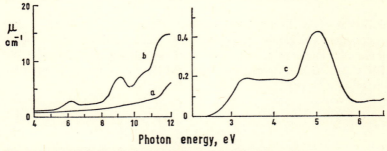

Fig. 9.14 Optical absorption bands produced in LiF : Mg.

give absorption at 310 nm and probably are variants of the same centre. Peaks 3, 4 and 5 are certainly the result of electron untrapping, and glow peaks 2 and 3 come from defects related to those giving glow peaks 4 and 5.

A second type of experiment has been made on the changes in the dipole absorption or low angle X-ray scattering during the annealing stages. These changes have also been correlated with the five glow peaks produced by subsequent irradiation of the heat-treated sample[26]. Two dominant defects shown up by these techniques are the dipole formed by a substitutional magnesium ion adjacent to a lithium ion vacancy and a trimer formed in the (111) plane of lithium by a ring of three of the dipoles. Models for the centres are shown in Fig. 9.15. The dipole and trimer are in dynamic equilibrium with reaction rates governed by the temperature, so that a high-temperature anneal, say 400°C, causes the trimers to dissociate. Alternatively prolonged annealing at a lower temperature, say 100°C, results in the

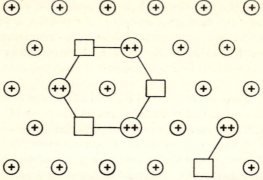

Fig. 9.15 A model of the dipole and trimer formed by Mg impurities linked to Li vacancies in the (111) plane of LiF.

diffusion of dipoles to form trimers and these may then form higher aggregates. The results of Fig. 9.16 by Dryden and Shuter[26] follow these two steps by measurements of the dielectric absorption. At 101°C only trimer formation occurs for the first 100 hours, but after

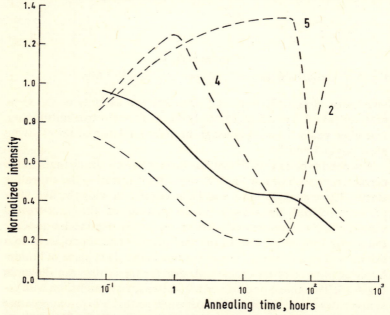

Fig. 9.16 The effect of annealing LiF : Mg at 101°C on the dipole concentration (——) and the glow peaks number 2, 4 and 5.

this shoulder the aggregates are forming. The parallel measurements of the thermoluminescence peaks suggest that peak 2 is related to an electron trapped near a dipole and that peaks 4 and 5 are in some way associated with trimers. This interpretation is very speculative, and when one considers all the annealing data the only clear results are that peaks 2 and 5 move in opposite senses with heat treatment and peaks 4 and 5 have some common features.

The interpretation of the role of dipoles and trimers is consistent with the need for both a quench from 400°C to preserve the magnesium in a dispersed state, and a low-temperature anneal at 100°C to form trimers.

The role of the other impurities, particularly the titanium, is not known, but it may influence the efficiency of recombination rather than the trapping site. Over all, the luminescence efficiency, that is, photons produced per released electron, is only about 1%, and the efficiency of storage of radiation-released electrons is about 0·05%. Both parameters can be improved by a better understanding of the defect reactions which are involved.

9.9 Archaeological and geological applications of thermoluminescence

The understanding of the properties of defects in insulating materials can be applied in many fields, and not the least of these are the fields of archaeology and geology. Although the geologists may not have required any knowledge of solid state theory, they have used some intuitive or empirical defect results in their work. It is obvious that to a prospecting mineralogist the absorption bands of sapphire or ruby are sufficient to identify the stones, and the understanding of their common crystal structure and the role of impurities is unnecessary. There is the possibility that with more refined techniques the slight changes in impurity levels and crystal properties that occur as one traverses a mineral deposit might be used to indicate the direction and concentration of a more valuable seam of minerals. For example, the association of gold and quartz is common, and gold may exist as a defect in the quartz. This association was eventually appreciated in the early attempts to produce high-purity germanium by zone refining. The purification which was carried out in quartz tubes actually contaminated the germanium with gold by ion exchange from the quartz, even though the gold was only present in the quartz tube in a few parts per million.

Archaeological applications also appear possible, if one considers

CCS—P

the absorption bands in glassware, since these may be characteristic of the sand used in some particular era and locality. The preceding examples are speculative, but the following example of using thermoluminescence measurements to date both rocks and pottery has been successfully applied and is becoming a valuable tool in archaeology[27, 28, 29].

The manufacture of pottery is an ideal example of sample preparation for thermoluminescence study, because the clay must be fired in a kiln at a temperature of 700–1000°C. This totally empties all the shallow electron traps. Further, the material is both a thermal and electrical insulator, so must be cooled slowly to avoid thermal stress and cracking. In so doing one ensures electrical equilibrium and the trapping levels remain empty. From this point onwards the pottery contains a 'clock' in the form of electron traps which are being progressively filled by internal radiation from radioactive decay, or, for buried material, from the gamma emission from the soil. To read the clock we must count both the number of filled traps and the rate at which they are filling during the radiation. The trap population can be detected by the strengths of the glow peaks, but for the rate of filling we need a careful analysis of the concentration of the various radioactive ions in the pottery. The analysis reveals the type of ion, the decay scheme and the mode of decay. To determine the efficiency of the various types of irradiation on the rate of trap filling, we can perform laboratory irradiations with controlled doses, which will also produce thermoluminescence signals.

Typical pottery clays contain radioactive material in the form of ^{238}U and ^{232}Th in concentrations of p.p.m. and radioactive ^{40}K up to hundreds of p.p.m. These isotopes have very long half-lives so the radiation flux is essentially constant throughout the life of any pottery and is of the order of 1 rad per year (i.e. 10^{-2} joules absorbed per kg per year).

It was apparent in Chapters 3 and 4 that the filling of electron traps is a very complex function of the trap distribution, type of ionizing radiation and irradiation flux. Also that in order to observe the glow peaks there must be suitable fluorescent levels, and the 'shallow' traps must not be so shallow that they are thermally emptied by long periods at ambient temperatures. We can now treat each of these problems in turn. The first correction is to account for the different ionizing efficiencies of the various types of irradiation, both internal and from the surrounding soil. Aitken[28] quotes the relative contributions to the dose from α-, β- and γ-radiation for a typical buried vessel as 57%, 24% and 19% respectively. The soil correction may sometimes be measured by placing a known sample in the location for a prolonged period. Secondly we hope that there is no flux dependence, because we wish to compare the natural rate for trap filling with the laboratory calibration experiments where the flux may be 10^4 greater. Fortunately this seems a valid assumption for pottery, and in certain cases it has been substantiated[23]. This is not necessarily in conflict with the flux-dependent alkali halide results described in § 3.7, because for pottery we are considering the total light output from a fixed number of traps which are mostly impurity centres and are not concerned with the production of new intrinsic defect centres.

Thirdly we shall only consider high-temperature peaks in the range 300 to 500°C, because the lower temperature peaks will be slowly drained at room temperature and the higher peaks will be masked by the thermal background radiation. To a certain extent this can be avoided if we analyse the emission spectrum from each glow peak or limit the detection to higher energy photons (blue or ultra-violet region).

To make quantitative measurements the pottery samples should be taken from interior sections of the vessel and then ground in a mill into powder, without exposing them to light[30]. This will avoid bleaching effects of sunlight or room light on the outer layer of the specimen. For buried samples there are also problems of a variable water content which can be as much as 10% by weight and so alter the effective radiation flux.

Allowing for such factors the current accuracy for dating by this technique is $\pm$ 10%. It seems improbable that it could ever be better than $\pm$ 5%. For pottery of the Romano-British era the date of firing is quoted within $\pm$ 80 years. Older specimens have also been used, and the results agree favourably with measurements of radiocarbon dating. An example of an Upper Paleolithic clay fragment from Dolni Vestonici has been dated as 31,000 $\pm$ 3000 B.C. by thermo-luminescence or at 28,500 B.C. by radioactive dating.

Variations on this theme are numerous, and one may list the possibilities of dating fossil shells (not previously annealed), dating geomagnetic field reversals in volcanic ash or measuring the thermal history of pottery or rocks.

Lava flow contains both magnetic and radioactive material, so on solidifying it both records the direction of the earth's field and also starts the 'clock' for a subsequent dating experiment. Such records are useful in determining the wanderings of the earth's magnetic poles, but the method is not applicable to the more spectacular reversals of magnetic field, since these occur in very much older samples of at least 700,000 years. Trap saturation and chemical effects on the lava would make the thermoluminescence dating less reliable on this large time scale. Lava sample studies by Aitken and Fleming[28] indicate that the dating method of thermoluminescence is possible for material up to at least 70,000 years old.

At the other end of the time scale the glow curve measurements provide a convenient method for checking the authenticity of doubtful artefacts. An excellent example is provided by Fleming et al.[31] in the study of Etruscan wall paintings on terracotta. While some samples correctly showed an age of 2000 years, many other samples were made in the last 20 years.

Thermal history studies also become possible if more than one glow peak exists in a sample. For example, a pot which had glow peaks at, say, 300 and 500°C would contain both peaks if it were kept at room temperature. However, if the pot were in a fire then the lower energy peak would be removed. In this way we might distinguish between cooking vessels and other artefacts. Similarly the relative heights of glow peaks in rocks at various depths below ground level have been used to study the thermal history of rocks which were exposed to hot sunlight, and to decide whether archaeological pits were used for cooking or storage.

9.10 Some other applications of defects

To conclude our list of applications we will briefly mention a few further topics.

Chemical analysis of materials by characteristic luminescence, thermoluminescence or ESR has the advantages that not only does one have a high sensitivity, but also the analysis is non-destructive. In the case of ESR there is the added benefit that one actually knows the defect site for some of the impurity ions. All three techniques have the shortcoming that one only senses impurities at particular sites and does not measure the total impurity concentration. However, in the impurity range of parts per million the alternative analyses are also prone to error.

Electro-optic devices generally operate via bulk properties of the crystals, but since defects are unavoidable their effects must be considered. At high light levels permanent refractive index changes take place, as in the $LiNbO_3$ crystals used for holographic image storage. Also Faraday rotation of light beams is produced at some defect sites which may degrade the performance of the devices.

Photo-emissive surfaces such as those used in photomultiplier tubes also require sensitization to liberate electrons in a particular photon energy range and a low work function for the electrons to escape. The preparation of the surface is a particularly awkward problem to study, because the defects may be similar to bulk defects, but the concentration of surface sites is low so they are difficult to detect in the presence of bulk defects. ESR measurements of surface ions have been used to give the concentration of attached impurities, but rarely can one interpret the data to give a specific site. Theoretical calculations of surfaces are also suspect, and frequently the same authors give opposite results depending on their choice of theoretical model. A further surface effect is seen in secondary electron conduction devices. Here the material is made in a sintered form with large internal spaces. Secondary electron emission in these materials is channelled in the spaces of the sinter, and collisions with the walls produce a gain in the secondary electron yield. The 'defects' are macroscopic, but the role of impurities in the host material is similar to that for the photocathodes.

Defects in reactor fuels are of great industrial consequence, and here the studies tend to explain why and what type of defects are formed, the way they cluster and hence the type of deformation that occurs in the material. The aim is to develop a doping method or an *in situ* annealing cycle, over an accessible temperature range, which removes the radiation damage without leaving stable defects behind.

9.11 The preparation of pure crystals

Throughout all our discussions of defect properties in insulators and the appearance of the defects during irradiation, we have emphasized the need to work with materials of extremely high purity and low intrinsic defect concentrations. We should realize that the more sensitive measurements of thermoluminescence, ESR or even optical absorption can detect as few as 10^{15} centres cm^{-3}, in other words atomic concentrations of parts in 10^7. Crystals obtained from commercial suppliers are rarely, if ever, of the quality necessary for measurements of this precision. It is not a reflection on the ability of the manufacturers, since the bulk of their material is not intended for laboratory use. However, it does mean that unless some industrial need generates high-purity material, as with silicon, there is no suitable source of crystals for research and the experimentalist must produce his own material. To demonstrate that such a task is both feasible and worth while we shall relate the approach used by Butler *et al.*[32] at the Oak Ridge Laboratory to produce large, pure, single crystals of KCl. The specific chemistry required for KCl is unimportant, but the overall approach has many features which have more general significance. Their success is measured by their ability to start with reagent grade material and yet produce single crystals containing less than 1 $\mu g/g$ of foreign alkali or halide elements, and also reduce the level of OH and heterovalent cations below 10 and 300 p.p.b.

The Oak Ridge approach has been to raise the purity level for all elements and still produce large single crystals. An alternative treatment might have been to use zone refining techniques to eliminate a particular element at the expense of retaining others. Unfortunately the zone refining is not only selective, so that the total impurity level might be high, but also the steep temperature gradients needed result in relatively small single crystals which are badly strained.

Before attempting any purification Butler *et al.* analysed the reagent grade KCl powder that they were to use and rejected batches of material that they thought to be unacceptable. One reason for this is that the major impurities in KCl are other alkali halides such as Na, Rb, Br and I which occur in concentrations up to 300 $\mu g/g$. Chemical treatments cannot eliminate these impurities but only reduce them, so it is essential to start with the highest grade material. Typical analyses of the reagent grade powder are given in Table 9.3.

The major steps in the preparation of pure KCl were an extensive purification of the KCl powder, followed by crystal growth from the melt on to a seed crystal which is in an inert atmosphere. The chemical steps of the purification and improvement produced at each stage are shown by Fig. 9.17.

Butler *et al.* have emphasized that one must pay meticulous attention to detail throughout the purification, otherwise fresh impurities will be introduced. Throughout the treatment the materials were handled under clean room conditions in a dust-free atmosphere and the personnel wore lint-free clothing. The analysis of the reagent KCl and chemicals used in the preparation obviously helps to maintain this standard. Further precautions were required, for instance, to remove boron introduced from the Pyrex ware; methanol was added to form methyl borate; the silica vessels were intensively outgassed to remove absorbed water; the thermocouple was placed outside the

Table 9.3 Common impurities in reagent grade KCl powder. µg/g.

	Ba	Br	Ca	Cs	Fe	I	Li	Mg	N	Na	P	Pb	Rb	S	Si	Sr
Lot I	<10	220	<3	<2	<1	13	<1	<2	20	18	<2					<3
Lot II	5	341	<1	<1	<0·3	4	<0·1	<3	<1	43	<0·5	<0·3	60	<1	4	<1
Lot III	<1	236	<1	<1	<0·3	2·2		2	4	5	<0·5	<0·5	7	<1	<1	<1
Lot IV	<1	167	<1	<1	<0·3	2·3	<0·1	2	<2	14	<1	1·5	9	<1	<2	<1
Lot V		189	<4	2	<0·5	3·4		<3	3·4	16	<1	<1	5	4	27·6	<1

CHEMICAL PURIFICATION

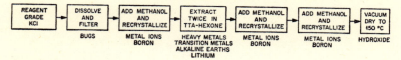

CHEMICAL PURIFICATION flow: REAGENT GRADE KCl → DISSOLVE AND FILTER (BUGS) → ADD METHANOL AND RECRYSTALLIZE (METAL IONS BORON) → EXTRACT TWICE IN TTA-HEXONE (HEAVY METALS TRANSITION METALS ALKALINE EARTHS LITHIUM) → ADD METHANOL AND RECRYSTALLIZE (METAL IONS BORON) → ADD METHANOL AND RECRYSTALLIZE (METAL IONS BORON) → VACUUM DRY TO 150 °C (HYDROXIDE)

FUSION-FILTRATION

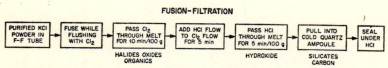

FUSION-FILTRATION flow: PURIFIED KCl POWDER IN F-F TUBE → FUSE WHILE FLUSHING WITH Cl₂ → PASS Cl₂ THROUGH MELT FOR 10 min/100 g (HALIDES OXIDES ORGANICS) → ADD HCl FLOW TO Cl₂ FLOW FOR 5 min → PASS HCl THROUGH MELT FOR 5 min/100 g (HYDROXIDE) → PULL INTO COLD QUARTZ AMPOULE (SILICATES CARBON) → SEAL UNDER HCl

Fig. 9.17 Summary of the chemical procedures used to purify and grow KCl crystals (after Butler *et al.*[32]).

melt. The molten salt was at one stage filtered through a glass frit into a silica storage bulb. The bulb was cooled with dry ice, because at that temperature the salt does not react with the silica. Failure to do this results in a strong bonding action which removes silica fragments into the purified powder. Similar precautions are given by Butler *et al.* for other stages of the preparation.

At the final impurity concentrations that one hopes to achieve the detection of specific impurities in any quantitative manner becomes

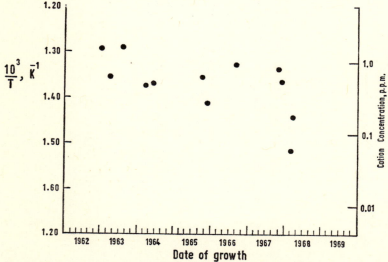

Fig. 9.18 The electrical conductivity 'knee' of KCl prepared by Butler *et al.*[32]. The later material contains less heterovalent cation impurities.

very difficult, as standard analysis techniques have a lower sensitivity limit of parts per million for many elements. The measurement of ionic conductivity is one convenient test of the total heterovalent cation impurity concentration. The temperature of the 'knee' in the plot of conductivity versus temperature is also a measure of the total impurity concentration, as this break in slope indicates the change in conductivity from intrinsic to extrinsic conduction. The progress of the Oak Ridge group can be measured by the improvement in knee temperature or by direct analysis for specific impurities. Figs 9.18 and 9.19 show that the changes introduced over the period 1962–1969

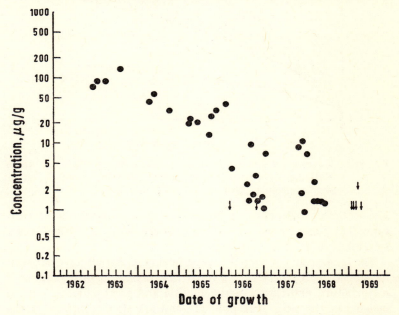

Fig. 9.19 The reduction of Bromine in KCl versus the date of crystal growth (after Butler *et al.*). Arrows represent a limit of sensitivity.

produced a systematic improvement in the quality of the final crystal. In so far as these changes are related to colour centre formation, we might note that the sensitivity of first stage coloration of the F band (§ 3.7) is a function of the impurity content, so that the reproducibility of the Oak Ridge crystals will be demonstrated by the gamma-ray induced growth curves of the F centres. A comparison of growth curves is shown in Fig. 9.20 for samples obtained from several suppliers. The extreme limits differ markedly, although the 19 Oak Ridge National Laboratory crystals all showed colouring curves which fell in the shaded region.

The problem of making crystals of the perfection that one would

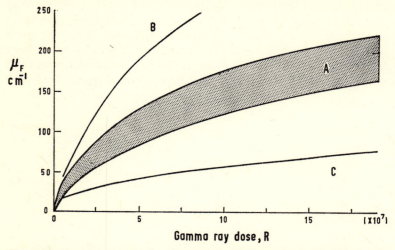

Fig. 9.20 F centre production in various KCl crystals. All Oak Ridge National Laboratory crystal results fall in region A; other sources produce data up to lines B or C.

like for defect studies has not been solved, but at least this work indicates that major improvements are possible when one undertakes a systematic method of purification.

The fundamental reason that experimental materials have not been developed to the necessary level is that crystal growth requires a major effort, first-class experimentation and great patience, but the benefits of the work lack publicity and are slow to appear. Consequently there are few scientists prepared to make this effort.

9.12 Conclusion

Our list of examples of the way in which defects control the properties of materials is extensive and includes such major fields as semiconductors, photography and luminescence. The most surprising fact that emerges is that the role of the defect or impurity in the majority of cases is not fully understood, yet industrial applications have progressed surprisingly well without this knowledge. However, the intuitive and empirical approach which has been so successful is based on the detailed results of the successful case studies of analogous systems. This parallel development of detailed research and empirical applications has been very fruitful, and we are confident that the development of new materials and applications will progress for many years to come.

Chapter 9 References

[1] Bube, R. H., *Photoconductivity of Solids* (Wiley) 1960.

[2] Dearnaley, G., *Contemp. Phys.* **8**, 607, 1967.

[3] Jackson, G., *Opt. Acta.* **16**, 1, 1969.

[4] Kiss, Z. J., *Phys. Today* **23**, 42, 1970.

[5] Taylor, M. J., *Phys. Bull.* **21**, 485, 1970.

[6] Dorion, G. H. and Roth, W., paper presented at the eleventh technical meeting of AGARD panel on displays for command and control, 1966.

[7] Smith, G. P., *I.E.E.E. Spectrum* **3**, 39, 1966.

[8] Berezhnoy, A. I., Gel'berger, A. M., Gorbachev, A. A., Piterskikh, S. E., Polukhin, Yu. M. and Yusim, L. M., *Sov. J. Opt. Tech.* **36**, 616, 1969.

[9] Staebler, D. L. and Schnatterly, S. E., *Phys. Rev.* **B3**, 516, 1971.

[10] Taylor, M. J., *Phys. Lett.* **A27**, 32, 1968.

[11] Faughnan, B. W. and Kiss, Z. J., *Phys. Rev. Lett.* **21**, 1331, 1968.

[12] Amodei, J. J. and Bosomworth, D. R., *Appl. Optics* **8**, 2473, 1969.

[13] Chen, F. S., LaMacchia, J. T. and Fraser, D. B., *Appl. Phys. Lett.* **13**, 223, 1968.

[14] Schneider, J., Marrone, M. and Kabler, M. N., *Appl. Optics* **9**, 1163, 1970.

[15] Rosenthal, A. H., *Proc. I.R.E.* **28**, 203, 1940.

[16] King, P. G. R. and Gittins, J. F., *I.E.E. Journal* **93**, 822, 1946.

[17] Stookey, S. D., *Ind. Eng. Chem.* **45**, 115, 1953.

[18] Parker, C. J., *Appl. Optics* **7**, 735, 1968.

[19] Leverenz, H. W., *An Introduction to the Luminescence of Solids* (Dover) 1968.

[20] Goldberg, P., *Luminescence of Inorganic Solids* (Academic Press) 1966.

[21] Henisch, H. K., *Electroluminescence* (Pergamon) 1962.

[22] Zimmerman, D. W., Rhyner, C. R. and Cameron, J. R., *Health Phys.* **12**, 525, 1966.

[23] Tochilin, E. and Goldstein, N., *Health Phys.* **12**, 1705, 1966.

[24] Mayhugh, M. R., Christy, R. W. and Johnson, N. M., *J. Appl. Phys.* **41**, 2968, 1970.

[25] Mayhugh, M. R., *J. Appl. Phys.* **41**, 4776, 1970.

[26] Dryden, S. and Shuter, B., to be published.

[27] Brothwell, D. and Higgs, E., eds. *Science in Archaeology* (Basic Books, N.Y.) 1963.

[28] Aitken, M. J., *Rep. Prog. Phys.* **33**, 941, 1970.

[29] McDougall, D. J., ed. *Thermoluminescence of Geological Materials* (Academic Press) 1968.

[30] Fleming, S. J., *Archaeometry* **12**, 133, 1970.

[31] Fleming, S. J., Jucker, H. and Riederer, J., *Archaeometry* **13**, 143, 1971.

[32] Butler, C. T., Russell, J. R., Quincy, R. B. and La Ville D. E., *J. Chem. Phys.* **45**, 968, 1966.

INDEX